THE AFRICAN-AMERICAN SOLDIER

Other books by Michael Lee Lanning

The Only War We Had: A Platoon Leader's Journal of Vietnam
Vietnam 1969–1970: A Company Commander's Journal
Inside the LRRPS: Rangers in Vietnam
Inside Force Recon: Recon Marines in Vietnam (with Ray Stubbe)
The Battles of Peace
Inside the VC and NVA (with Dan Cragg)
Vietnam at the Movies
Senseless Secrets: The Failures of U.S. Military Intelligence
The Military 100: A Ranking of the Most Influential Military Leaders of All Time

THE AFRICAN-AMERICAN SOLDIER

From Crispus Attucks to Colin Powell

Lt. Col. (Ret.) Michael Lee Lanning

A Birch Lane Press Book
Published by Carol Publishing Group

A Birch Lane Press Book
Published by Carol Publishing Group
Birch Lane Press is a registered trademark of Carol Communications, Inc.

Editorial, sales and distribution, rights and permissions inquiries should be addressed to Carol Publishing Group, 120 Enterprise Avenue, Secaucus, N.J. 07094

In Canada: Canadian Manda Group, One Atlantic Avenue, Suite 105, Toronto, Ontario M6K 3E7

Carol Publishing Group books may be purchased in bulk at special discounts for sales promotion, fund-raising, or educational purposes. Special editions can be created to specifications. For details, contact Special Sales Department, 120 Enterprise Avenue, Secaucus, N.J. 07094.

Manufactured in the United States of America
10 9 8 7 6 5 4 3 2

Library of Congress Cataloging-in-Publication Data
The African-American Soldier : From Crispus Attucks to Colin Powell /
by Michael Lee Lanning.
p. cm.
"Birch Lane Press book."
Includes bibliographical references and index.
ISBN 1-55972-404-8 (hardcover)
1. United States—Armed Forces—Afro-Americans—History.
I. Lanning, Michael Lee.
UB418.A47A35 1997
355'.008996'073—dc21 96-37626
CIP

To

James Walter Lanning

and

Judy Edwards Lanning

CONTENTS

ACKNOWLEDGMENTS

Assisting in research for this book were the staffs of the National Archives and Records Administration in Washington, D.C.; Office of Public Affairs, U.S. Army; Office of Information, U.S. Navy; Public Affairs Office, U.S. Marine Corps; Public Affairs Office, U.S. Air Force; and the Public Affairs Offices of the U.S. Military, Naval, and Air Force Academies.

Providing library resources were the Library of Congress, Washington, D.C.; the Pentagon Library, Department of Defense; the Arizona State University Library, Tempe, Arizona; and the public libraries of Phoenix, Scottsdale, and Tempe, Arizona.

Aiding in the acquisition of special reports were the staffs of Arizona congressman Ed Pastor, California congressman Ronald V. Dellums, and the Department of Defense Office of Equal Opportunity.

Assisting in photographic research were JoAnna M. McDonald of Carlisle, Pennsylvania; Mary Beth Straight Kiss of the U.S. Naval Institute; and the staff of the Library of Congress.

Offering research support, advice, and encouragement on a regular basis was my good friend Dan Cragg of Springfield, Virginia.

Thanks to Hillel Black of the Carol Publishing Group and to my agent, Russell Galen, for bringing it together.

A special thanks goes to Dr. Dudley Taylor Cornish for introducing me to black military history.

Lastly, this book would not have been possible if not for the dedicated, long hours of editing and other assistance provided by Linda Ann Moore Lanning.

1

COLONIAL DAYS AND THE REVOLUTIONARY WAR

The year was 1770. Although it was six years before the American Revolutionary War, Boston colonists were already chafing under the presence of the garrison of British soldiers stationed by the Crown in their port city. The lingering winter weather and the arrogance of the redcoat soldiers made tempers short and the desire for change strong.

Warming himself near a fire and listening to the litany of complaints from his companions on the early evening of March 5, Crispus Attucks, too, was ready for action. As a runaway slave who had eluded capture for twenty years, Attucks knew firsthand about the unfair treatment of which the colonists spoke and the acts of rebellion about which they dreamed. At forty-seven, he was a veteran of both, a man who had escaped bondage to find his freedom on the hazardous and arduous high seas. Awaiting a break in the weather before his ship could sail, Attucks understood the risks of defiance better than most of those who were, in the comfort of the fire's warming radiance, recounting the British injustices against the Americans.

Suddenly, the group became aware of excited voices and the ringing of the town's alarm bell. Attucks was the first out the door to investigate the commotion. In the snow-covered streets he saw colonists clustered near the British garrison. In loud voices and with animated movements, the men seemed to be arguing among themselves, some pointing at the British, others trying to pull them away. As Attucks moved toward the center of the disturbance on King Street, some of the American colonists began to flee.

Attucks stopped the first man he met. In an agitated voice the colonist told Attucks that a British soldier had refused to pay a barber after receiving a haircut. The young man had pursued the soldier into the street, demanding his money. The soldier laughed and knocked the youth to the ground. Outraged witnesses came to the support of the barber, only to have more soldiers arrive and threaten to kill them all. Incensed at the audacity of the threat, the colonists had begun verbally attacking the soldiers as they marched back to their barracks. The jeering crowd had grown but was undecided as to what to do. As Attucks watched, the colonists began reluctantly to retreat.

Angered by the story and the attitude of the British, the short, stocky Attucks advanced toward the garrison, determinedly pushing his way through the hesitant crowd and urging them to stand against the soldiers. With a leader at their helm, the colonists turned once more and followed Attucks to the customhouse. The crowd swelled and began hurling snowballs and rocks at the sentinels, who, terrified by the crowd's mood, called for reinforcements.

With the colonists behind him, Attucks led the confrontation, throwing snowballs and anything else he could find on the street at the concentrated redcoats. At one point, he picked up a stick, which he menacingly wielded like a club as he dared the British soldiers to fire. Spurred on by the bravery of the curly-haired African American, the colonists unleashed even more obscenities and missiles against the soldiers.

The disciplined customhouse guards endured the chaos, knowing they could shoot only on orders from their officers or from the civil magistrate. Even the arrival of their captain, Thomas Preston, who tried in vain to disperse the crowd, did not quell the flurry of thrown objects and verbal abuse. Lack of action by the British served only to inflame the Bostonians further.

With at least a dozen men behind him, Attucks confronted the soldiers by striking their muskets with his stick. When the British captain stepped forward, Attucks swung his club at the officer's head. In warding off the blow, Preston bumped into one of his soldiers, causing him to drop his weapon.

Instantly, Attucks seized the musket. The soldier also grabbed for the gun, and the two struggled for control. In wresting the weapon from Attucks, the soldier stumbled and fell. While he was down, the colonists, now confident that the British would do nothing, turned their chants to "Why don't you fire? Why don't you fire?" Attucks, leaning on his stick, mockingly stared at the fallen redcoat. Humiliated, the British soldier regained his feet and fired directly into Attucks. Instantly, the other soldiers also opened fire.

When Attucks fell dead from two bullets in his chest, he became the first American to die in the course of events which would lead to independence for the United States. The colonists declared Crispus Attucks a hero, for he took action against the British when others were willing only to talk; he stood against subjugation when others were willing to submit; he sacrificed his life for freedom when others were willing to live under hellish conditions.

The irony of his heroism was that as a black man, Crispus Attucks would not have enjoyed the benefits of American independence had he lived. Although he knew this, he was willing to give his life for the good of the country. In so doing, he joined a long line of other African Americans who would also sacrifice themselves for freedoms and benefits they as individuals could not enjoy.

African Americans have always played a significant role in the military history of the United States. Their participation in, and contributions to, pivotal events parallel those of other Americans even before the Revolutionary War. For almost five centuries, blacks have been an integral part of the development of the Americas, fighting and sometimes dying in pursuit of a better life. In the case of the United States, African Americans have participated in every American conflict, often forced to struggle for the right to fight for their country. Ironically, like Crispus Attucks, many blacks have sacrificed themselves for the American cause in war while equality remained beyond their reach in peace, withheld by prejudice and ignorance.

The contributions made by African Americans in the New World began with the first European explorations. At least one man of African heritage, Pedro Alonso Niño, sailed with Christopher Columbus on his 1492 voyage of discovery. Another black sailor, Diego el Negro, accompanied Columbus on his last voyage to the New World in July 1502. In subsequent Spanish explorations, Africans served the

Crispus Attucks: the first casualty of the American Revolution in 1770. (LIBRARY OF CONGRESS)

military expeditions as servants, laborers, and soldiers. When Vasco Núñez de Balboa crossed Central America to reach the Pacific Ocean, at least thirty blacks occupied his ranks. Africans also accompanied Hernando Cortés in his conquest of the Aztecs in Mexico in 1519 and assisted Francisco Pizarro in conquering the Incas and Peru in 1533.

Enslaved blacks arrived in great numbers in the Western Hemisphere as the Spanish settled the Caribbean Islands, their first foothold in colonizing the Americas. It is not surprising that blacks were an integral part of subsequent Spanish conquests.

The first blacks to arrive in North America did so as a part of a brief effort by Spain to establish a colony near the mouth of the Pee Dee

River in present-day South Carolina. Details are sketchy, but records indicate that one hundred slaves accompanied five hundred Spaniards to establish a settlement in 1526. In a matter of only a few months, the effort failed, and the survivors returned to Spanish island communities in the Caribbean. Some accounts attribute the failure of the colony to an uprising by the slaves, but no factual information remains to substantiate this claim.

The most significant black man in the early explorations of North America arrived via shipwreck on the southeast coast of what is now Texas in 1528. Among the survivors were the expedition's leader, Cabeza de Vaca, the Spanish soldier Dorentes, and his black slave Esteban, who quickly became "twice a slave" when local Indians captured all the castaways. Eventually, Esteban, de Vaca, Dorentes, and several others escaped overland to Mexico City, arriving there after a journey of eight years.

Esteban remained a slave, but his skills in Indian sign language, learned while a captive, led to his earning another "first" in the New World. In 1536, Esteban, acting as a scout and interpreter for expeditions into the American Southwest, became the first non-Indian to visit what would become New Mexico and Arizona. Three years later, Esteban returned to the Southwest as a guide for Spaniards seeking the fabled "seven golden Cities of Cibola." Esteban found no gold but did reach the Zuñi village of Hawikuh in New Mexico, where the Indians, angered at the intrusion, executed him.

Unlike the Spanish, who had developed a slave system in the Caribbean, English settlers establishing colonies along the East Coast brought no Africans—slave or free—with them. Not until a little more than a decade after the first English settlement in America at Jamestown did the first Africans arrive among the English. Slaves joined the colony on August 20, 1619, when a Dutch ship anchored at the Virginia colony and offered twenty individuals, recently captured in Africa, in exchange for provisions.

Initially, the small population of enslaved blacks represented no threat to the white majority. In fact, the number of Africans in the North American colonies increased so slowly that it was not until 1638 that the first slaves appeared on the auction block in the Massachusetts Bay Colony. The blacks provided what the whites perceived as reinforcements from the real threat to the colonies—the Native American Indians. During this time the colonial militias welcomed both free and slave Africans into their ranks to repel Native American attacks and to join the occasional offensive operations against Indian villages.

As the slave population increased, however, the subjugated blacks, many of whom were now trained in arms and military procedures, themselves became a threat to the whites. In 1639 the General Assembly of Virginia passed the first act in American history excluding blacks from military service, declaring, "All persons except Negroes to be provided with arms and ammunition or be fined at pleasure of the Governor and Council."

In 1656, Massachusetts followed Virginia's lead and excluded blacks from service in the colonial militia. Connecticut did likewise after a joint uprising of Indians and blacks near Hartford in 1661. By the end of the seventeenth century, all of the colonies had officially excluded blacks from serving in the military. Some of them did amend the restrictions and allowed free blacks and a few slaves to serve as drummers, fifers, cooks, and laborers as well as in other unarmed positions.

In terms of self-protection, the whites' precautions were not without foundation, for the blacks did not accept their bondage peacefully or willingly. In 1663 the first recorded slave uprising in the English colonies occurred in Gloucester County, Virginia. For the next hundred years, more than 250 slave revolts occurred throughout the colonies. The whites routinely put down the rebellions by ruthlessly torturing and killing the instigators. Except for a few African communities, whose citizens, known as Maroons, maintained their freedom in remote mountain or swamp villages, no attempts at gaining freedom were successful.

Despite their potential threat to the whites, blacks did fight in the militias during the various conflicts with the Indians, and later against the French, in the first half of the eighteenth century, when white colonists either rescinded or ignored the various laws restricting African Americans from active military service. As would be the model for centuries to come, blacks served in armed positions when white Americans felt threatened by outside enemies, fighting and dying for a culture which did not recognize them as equals.

Few of the accomplishments of black Americans of the time found their way into recorded history, but what records do exist show that blacks served loyally and well. Blacks fought and died in the colonial militia in King William's War (1689), Queen Anne's War (1702–13), and in numerous campaigns against the Indians of various magnitude. In some cases, slaves who distinguished themselves in battle became free men. During fights against Indians in 1703, South Carolina offered emancipation to any slave who killed or captured one or more of the hostiles. The state, "at the charge of the public," compensated the slaves' owners. In most instances, however, at the conclusion of the crisis, black soldiers found themselves once more disarmed and returned to their masters.

During the Seven Years' War of 1756–63 blacks joined the colonial militias in support of the British against the French and their Indian allies. Once again, most blacks served in support positions as laborers and wagoners; only a few acted as scouts and regular soldiers. The slaves received pay equal to that of their fellow white soldiers but had to surrender all or part of the wages to their owners. As in the past, some blacks gained their freedom for honorable service, but the majority of the slave-soldiers returned to bondage at the end of the hostilities.

The only place where any degree of real racial equality existed in pre–Revolutionary War America was at sea. Again, circumstances rather than any compassion among the white majority were responsible. Conditions aboard merchant and fishing vessels as well as on military privateers were so intolerable that many white sailors deserted. Free blacks and runaway slaves welcomed the opportunity for employment regardless of the conditions and shared mostly equal pay as part of integrated crews.

By 1775 the total population of the English colonies in America was about 3 million, 600,000 of whom were Africans or of African descent. In the southern colonies, where agricultural field crops benefited the most from indentured labor, slaves constituted almost 40 percent of the population. With so many held in bondage, the potential of a slave rebellion kept the white population so fearful that few African Americans received any military training.

Except for the threat of a slave uprising and the occasional use of black Americans in the militia when the situation warranted, white Americans appeared to pay little attention to the overall question of slavery other than its favorable impact on the economic development of the colonies and the enrichment of the colonists. Not until they themselves began to feel oppressed by increased control and taxation by Great Britain did some Americans begin to question their own position as oppressors and the morality of owning fellow human beings.

Despite their harsh treatment and continued enslavement, many African Americans "closed ranks"—as did Crispus Attucks—with the white colonists as the Revolutionary War neared. When the smoke cleared on Boston's King Street that March day in 1770, Attucks and four white patriots lay dead. The Boston Massacre became a rallying point for those opposing British rule and an anniversary marked by ceremony both during and after the Revolutionary War. In 1889 the city dedicated the Crispus Attucks Monument on Boston Common.

Despite the death of Attucks and the service of slaves in the militias, the efforts and stated purpose to free the colonies from Great Britain provided neither freedom nor equality for African Americans. Initial drafts of the Declaration of Independence in 1775 contained language damning slavery and extending freedom to all. However, because of objections from the southern delegations, the final version, which stated that all men are created equal and guaranteed "certain unalienable rights . . . life, liberty, and the pursuit of happiness" to all, did not apply to enslaved or free African Americans.

Although they faced exclusion from the objectives of the American Revolution, blacks participated in the war from its very beginnings. Several black militiamen were among those firing the "shot[s] heard round the world" at Lexington and Concord on April 19, 1775. One of the first Americans to fall at Lexington Green was Price Estabrook, a black soldier in Capt. John Parker's company of Minutemen.

With the Revolutionary War now officially begun, the Committee of Safety—also known as the Hancock and Warren Committee—which orchestrated the early efforts of the rebellion, met in May 1775 to determine the role of African Americans in the rebel military. The committee determined that while free blacks could serve, slaves could not because their service would be "inconsistent with the principles that are to be supported, and reflect dishonor on this Colony."

Blacks already in uniform remained on duty, however. While their numbers, both slave and freemen, were small, their bravery and impact were enormous. For example, some reports from the Battle of Bunker Hill on June 17, 1775, credit Peter Salem, a former slave freed shortly before the battle, with killing the British assault force commander Major John Pitcairn with a musket shot.

While Salem's accomplishments at Bunker Hill lack official substantiation, the brave service of fellow black soldier Salem Poor are well documented. Following the battle, Poor received praise from his white commanders when they reported that he "behaved like an experienced officer as well as an excellent soldier." He also received an official commendation for his performance in a report to the Massachusetts Bay General Court, dated December 5, 1775: "We would only beg leave to say, in the person of this said Negro centers a brave and gallant soldier."

Despite the number of black militiamen in the Revolution's early battles, African Americans had not yet officially gained the right to join the fight for American independence. On June 17, 1775, while the Battle of Bunker Hill raged in Boston, the Continental Congress assumed jurisdiction of the various colonial militias and formed the Continental army, with Gen. George Washington as commander in chief. Washington, a Virginia slave owner, who disliked the idea of black slaves or freemen serving with whites, excluded them from his ranks. On July 9, 1775, Washington's adjutant general issued orders to recruiters not to enlist "any deserter from the Ministerial (British) army, nor any stroller, Negro, or vagabond" in the Continental army.

During the next three months the numbers of the Continental army dwindled due to casualties, desertions, and completed enlistments; as a result, Washington and his staff met to reconsider the enrollment of blacks. Prejudices, however, continued to prevail over practicality. By a unanimous decision on October 8, Washington and his staff agreed to continue to exclude slaves from enlistment. By a large majority, they also concurred with the decision not to permit the enlistment of free blacks.

A few weeks later, the Continental Congress backed Washington and his staff and stated that blacks, free or slave, should be "rejected altogether" from the army. On November 12, 1775, Washington issued an official order preventing the enlistment of blacks and allowing those currently in uniform to remain only until their current enlistments expired. State and local militias across the newly declared United States of America followed suit.

The whites' reasons for denying blacks the right to fight for their country's freedom ranged from personal prejudice to economic considerations. Many American whites, particularly those in the South, considered blacks subhuman, inferior, and cowardly and refused to serve with them as equals. Although blacks fought bravely in all the early battles of the Revolution, white soldiers responded angrily to the taunts of the regular British soldiers, who shouted jingles about their African-American comrades:

> The rebel clowns, oh! what a sight
> Too awkward was the figure
> 'Twas yonder stood a pious wight
> And here and there a nigger.

White slave owners had no intention of giving up what they considered their private property for the war. They preferred to keep their "chattel" in the fields earning them a profit rather than risk their lives on the battlefield.

Not all Americans agreed with the policies of prohibiting blacks from the military. Freemen argued for the right to serve, and many slaves expressed their desire for the privilege to fight. Some officers within the army shared these sentiments. In an October 24, 1775, letter to John Adams, Gen. John Thomas wrote that he supported such enlistments because blacks "have proved themselves brave."

Whereas blacks faced exclusion from the white American militias, they found themselves welcomed into the British ranks for two reasons: Firstly, the British were happy to recruit African Americans to augment their personnel shortages that resulted from worldwide commitments of the empire. Secondly, the British sensed how divisive slavery was among the white Americans and wished to exploit the weakness.

On November 7, 1775, John Murray, the earl of Dunmore and the royal governor of Colonial Virginia, issued a proclamation, "I do hereby . . . declare all indentured servants, Negroes, or others (appertaining to the rebels) free, that are able and willing to bear arms, they joining His Majesty's troops, as soon as may be, for the more speedily reducing this Colony to proper dignity."

Within a month, three hundred escaped slaves joined Dunmore's Ethiopian Regiment, donning uniforms inscribed with the slogan Liberty to Slaves. Over the next few months more than thirty thousand slaves crossed the lines, seeking the freedom denied them by their American masters.

Other British units accepted runaway slaves in their camps as laborers, wagon masters, and horse handlers. Some became paid servants to officers, while others took more unusual jobs. The hangman at the 1776 execution of rebel spy Nathan Hale in New York was Bill Richmond, "a man of color."

Pressured by the loss of slaves to Lord Dunmore, a shortage of enlistees, and continuous protests by the free black community, Washington reconsidered his exclusion policy. On December 30, 1775, the American general issued orders that authorized the enlistment of free blacks but continued the ban on recruiting slaves. He reinforced his denial of slave enlistments in a general order on February 21, 1776.

Despite official policy, slaves did join the ranks of the Continental army. Sometimes they were able to enlist because subordinate unit commanders, always short of personnel, refused to quibble over the "free" or "slave" status of blacks and integrated the volunteers directly into the ranks. In other instances, slaves joined as "substitutes" for their masters, who chose not to personally fight for independence.

The situation changed when the British defeated Washington at New York in the fall of 1776 and the Continental army had to confront the fact that, with its numbers and morale in jeopardy, the success of the Revolution was in doubt. The turning point for black enlistments came in September when the Continental Congress asked that states raise eighty-eight battalions to reinforce the army, a request followed three months later by an order for an additional sixteen battalions, with instructions that the ranks be filled "by drafts, from their militia, or in any other way."

Black enlistments increased dramatically when some states, especially those in the northern and mid-Atlantic regions, liberally interpreted "in any other way" as authorization to recruit African Americans. These states enticed both freemen and slaves to enlist, offering slaves freedom at war's end. An additional incentive for securing slave enlistments was the guarantee to owners that the states would compensate them at market value for all freed slaves. Only in the Deep South did blacks face continued resistance to their enlistment.

By mid-1776 blacks served in nearly every battalion in the Continental army as either infantrymen in regiments recruited in the North or as servants and support personnel in southern regiments. Reports by British officers noted that most of the units they faced had several black faces interspersed in their ranks.

Whether volunteer or substitute, black soldiers acquitted themselves well. A Hessian mercenary officer in service of the British recorded in his personal journal on October 23, 1777: "The Negro can take the field instead of his master; and therefore no regiment is to be seen in which there are not Negroes in abundance; and among them are able-bodied, strong, and brave fellows."

Washington, now more concerned with the possibility of losing the war than with his objections to African Americans, also accepted their service. Several black soldiers, including Prince Whipple and Oliver Cromwell, accompanied Washington's night crossing of the Delaware on Christmas, 1776, to surprise and defeat the British-Hessian garrison at Trenton. Although not confirmed by recorded history, several of the later paintings of Washington crossing the icy river depict one or more black men aboard his boat.

Throughout the Continental army, officers recognized the need for black enlistees to bolster their ranks. Alexander Hamilton, a member of Washington's staff, wrote the president of the Continental Congress, John Jay, on March 12, 1779, declaring, "I have not the least doubt, that the Negroes will make very excellent soldiers."

Hamilton then continued with a thoughtful summary of the entire question of African-American enlistees, writing:

> The contempt we have been taught to entertain for the blacks, makes us fancy many things that are founded neither in reason nor experience; and an unwillingness to part with property of so valuable a kind will furnish a thousand arguments to show the impracticability or pernicious tendency of a scheme which requires such a sacrifice. But it should be considered, that if we do not make use of them in this way, the enemy probably will; and that the best way to counteract the temptations they will hold out will be to offer them ourselves. An essential part of the plan is to give them their freedom with their muskets. This will secure their fidelity, animate their courage, and I believe will have a good influence upon those who remain, by opening the door to their emancipation.

More and more freemen and slaves enlisted, and blacks served with whites in integrated units, but in limited roles—mostly as laborers and support personnel. But necessity soon provided new opportunities for black soldiers. Rhode Island, with a small population and two-thirds of its territory occupied by the British, experienced great difficulty in maintaining the number of able men in its two active-duty battalions. At Valley Forge in November 1779, Gen. James Varnum proposed to Washington that the two diminished units be combined into one and that officers return to Rhode Island to recruit an all-black battalion.

Both General Washington and Rhode Island governor Nicholas Cooke concurred, admitting that there was no other way for the state to meet draft quotas. When the Rhode Island Assembly passed a measure authorizing the battalion and granting payment to owners of enlistees, they made an effort to sound noble instead of reflecting on need: "History affords us frequent precedents of the wisest, freest, and bravest nations having liberated their slaves and enlisting them as soldiers to fight in defense of their country."

Five companies, totaling 226 men, commanded by white colonel Christopher Greene, initially composed Rhode Island's "black battalion." Shortly after their formation in August, the battalion participated with white units in the Battle of Rhode Island. Although they had not received proper infantry training, the black battalion held the line for four hours and suffered twenty-two casualties as they stood against assaults by British-Hessian infantry in what became their most significant battle.

Even though the citizens of Rhode Island were reluctant to relinquish their slaves to the war effort, the black battalion maintained an average strength of 150 for the remainder of the conflict. In addition to the August campaign in their own state, they fought in New Jersey and New York, where, at Point's Bridge in May 1781, Greene died from wounds received during a night attack. His replacement, Col. Jeremiah Olney, commanded the unit until war's end and hailed his troops for their "unexampled fortitude and patience."

Other states considered organizing all-black units, but except for the brief appearance by a company from Boston known as the Bucks, only Connecticut fielded another black unit for any length of time. The Connecticut Colonials company, composed of fifty-two freemen and slaves, served from June 1780 to November 1782 as an all-black unit before being disbanded, its members distributed among the white units for the final months of the war.

Allies of the newly declared United States employed their own black soldiers in support of the Revolution. Both France and Spain dispatched military expeditions to North America to fight their common enemy, Great Britain. In 1779 a French force of more than thirty-five hundred men joined American troops at Savannah, Georgia, in an attempt to liberate the city from its British occupiers. A legion of nearly six hundred black freemen and slaves, recruited in the West Indies and commanded by Viscount François de Fontanges, filled the ranks. One of the black soldiers, Henri Christophe, later played an important role in the slave uprising that forcibly took over the Caribbean island nation of Haiti.

Efforts by the French and Americans failed at Savannah, but the black soldiers performed well nonetheless. The official accounts of the battle sent to France commended the blacks, stating, "The legion saved the army at Savannah by bravely covering the retreat."

The Spanish governor of Louisiana, Bernardo de Galvez, also used black soldiers in his campaigns to drive the British from the Mississippi River valley. Estimates of the number of blacks, of every rank, in Galvez's army range from as low as 10 percent to as high as 50. Galvez's army pushed the British out of Louisiana and then continued their march along the Gulf Coast, securing Mobile and Pensacola. During the campaign, six black officers were cited for bravery, and the king of Spain awarded them medals of valor.

In addition to serving in the Continental army, state militias, and units of U.S. allies, African Americans fought in the new nation's navy. Conditions aboard ships had improved little since the early days of the colonies, and sailors willing to accept the dangers and hardships of the sea were difficult to find. As a result, the Continental navy welcomed black sailors on their man-of-wars and as crew members of state naval vessels and privateers.

Neither the states nor the federal government passed any legislation or issued any orders forbidding naval recruitment of blacks. As early as 1775 the various American navies actively recruited black seamen. A Newport, Rhode Island, recruiting poster boldly proclaimed: "Ye able backed sailors, men white or black, to volunteer for naval service in ye interest of freedom."

Black sailors saw action in every major naval battle of the Revolutionary War, including as a part of John Paul Jones's crew aboard the

Bonhomme Richard that defeated the *Serapis* in their epic battle on March 7, 1778. Although no black captained a warship during the Revolution, several had the important duty of "pilot" in maneuvering ships into port and through intercoastal waterways.

Even the southern states welcomed black pilots onto their ships, for as one official explained, they were "accustomed to the navigation of the river." Virginia, with the largest state navy in the South, led the way in employing black pilots. One of these men, Caesar, the slave of Carter Tarrant of Hampton, operated the wheel of the schooner *Patriot* when it captured the British brig *Fanny*, securing a great trove of stores and supplies. An act of the Virginia legislature on November 14, 1789, purchased Caesar's freedom in reward for his "meritorious service" in continuing to pilot the state's vessels for the duration of the war.

At least thirteen African Americans served in marine units during the war. Three were members of the Continental marines, and the others were among crews in the Connecticut, Massachusetts, and Pennsylvania navies. One of the black marines, John Martin, died in action aboard the brig *Reprisal* in 1777.

From the first shots of the Revolution that felled Crispus Attucks at the Boston Massacre in 1770 to the formal end of the war in 1783, black Americans contributed to the fight that gained America's independence. At the victory review following the British surrender at Yorktown in 1781, Baron Ludwig von Closen, a German serving with the French in support the of Americans, noted, "Three-quarters of the Rhode Island regiment consists of Negroes, and that regiment is the most neatly dressed, the best under arms, and the most precise in its maneuvers."

Records of the American Revolutionary War do not provide exact information as to the number of its black participants. Enlistment records, when kept, did not necessarily note the race of a soldier or sailor. However, estimates consistently place the number of black military veterans of the rebellion at about 5,000 of the total of 300,000 Americans who served during the war.

The best summary of their performance did not appear in print for nearly a century, and it was at a time when blacks in uniform again became a major issue in the United States. An article entitled "Negro Soldiers in the Revolution," in the September 26, 1863, issue of the *Army and Navy Journal*, proclaimed: "The record is clear, that from the beginning to the conclusion of the war of the Revolution, Negroes served in the Continental armies with intelligence, courage, and steadfastness; and that important results in several instances are directly traceable to their good conduct."

Another thousand blacks served in Lord Dunmore's Ethiopian Regiment and in other British units, and two-thirds of the thirty thousand who sought protection from the British appeared in various support roles during the conflict. Oddly, the British in their opposition to the Revolution produced more free blacks than did the victorious Americans. Despite American protest, the British evacuated the blacks who had supported them along with their army at war's end. Some resettled in Canada, and at least three thousand landed in Nova Scotia. In 1792 some twelve hundred sailed to West Africa to establish the colony of Sierra Leone.

The British, however, took the majority of their black supporters to colonies in the West Indies. While most landed as freemen, the British sold as many as a thousand back into slavery to English plantation owners.

Generally, black slaves who fought on the American side in the war did receive their freedom, as the states had promised. Many black veterans even shared in the land grants provided veterans. However, relatively few African Americans had the opportunity to serve, and their contributions received little recognition from empowered white Americans, who quickly forgot the critical role blacks had played in attaining the country's freedom.

The American Revolution did, however, result in some improvements for the black population of the country. Many Americans recognized that slavery was inconsistent with the freedoms for which they had fought and with the rights promised in the Constitution. Several of the northern states, including Massachusetts, New Hampshire, and Vermont, completely abolished slavery. Other northern states began measures to gradually free their slaves, and even southern states eased their manumission laws.

The vast majority of African Americans, however, continued to live in the Deep South, where slavery remained essentially unchanged. The few blacks who were free endured treatment as inferiors and found themselves familiar with "equality" as a word rather than an experience.

The American Revolutionary War set precedents for blacks and whites alike. For blacks, service in wartime could become a pathway toward enhancing their status and gaining additional opportunity. Unfortunately, the white majority also set a pattern—to look to African Americans only in time of great military need and then to ignore them and their accomplishments when peace resumed.

2

THE WAR OF 1812 AND BEYOND

Despite their loyal service in the American Revolution, African Americans found that the postwar period held no rewards for them. Slavery continued to flourish in the South, and discrimination and segregation marked the lives of African Americans in the North. Irish and German immigrants poured into the country, particularly in the northern states, and took many of the menial jobs traditionally held by free blacks. The military, which had been an avenue to freedom and status for blacks, now, in peacetime, handed them setbacks. Huge manpower reductions dramatically scaled down the size of the Regular Army and state militias. No longer needed to fill dangerous combat roles, black Americans soon found themselves excluded.

On May 8, 1792, Congress passed the Militia Act, which called for the enrollment of "each and every able-bodied white male citizen between the ages of 18 and 45." Although the act made no mention of blacks, either free or slave, most federal and state recruiters interpreted the wording to exclude blacks altogether. Paradoxically, only militias in the South ignored the "whites only" implication. In Georgia and South Carolina freemen continued to enlist as laborers, pioneers, and musicians, and some slaves served in units when plantation owners "hired out" their black property for military support duties. In North Carolina free blacks could enlist in the militia and serve in a variety of positions until 1812, when restrictions limited them to duties only as musicians.

Interestingly, while the uniformed military excluded blacks, the small civilian-operated War Department that administrated the army did not. Caesar Lloyd Cummings, a free black, was one of the department's six full-time employees. He served as clerk, messenger, doorkeeper, and janitor.

Limitations for blacks in the military continued when a congressional act, passed on July 11, 1798, formally authorized "establishing and organizing a marine corps" in which no Negroes, mulattoes, or Indians could enlist. In keeping with the white precedent of enlisting African Americans when desperate for support and ignoring them after the crisis had passed, the legislation overlooked the record of the thirteen blacks who had served as marines during the Revolutionary War. Nevertheless, the newly authorized Marine Corps followed the letter and spirit of the law, excluding blacks from their ranks for more than a century—until the manpower shortages of World War II once again required that whites seek black reinforcements.

Initially it appeared that the navy, too, would follow the lead of the army and the marines. Indeed, Secretary of the Navy Benjamin Stoddert echoed the marine order of "no Negroes, mulattoes, or Indians" to his recruiters in August 1798. However, to the blacks' advantage, the navy faced difficulties the other services did not. The army had a land force that required few soldiers, and there was little competition from other employers for their services. The navy, though, had to compete with merchant vessels and fishing fleets to enlist experienced sailors who could cope with the hardships and dangers of life on the sea.

As a result, blacks steadily found opportunities available in the navy. There is no evidence that Stoddert's order received much attention, and if a ban against recruitment of black sailors did ever occur, it was of short duration. In fact, during the time of the order and for several years afterward, names of black sailors appeared on nearly every U.S. Navy ship's crew list, including that of its primary warships the *Constitution* and the *Constellation*.

The status of black Americans did not improve with the arrival of the nineteenth century. Except for the navy, service in uniform was prohibited. Even the Louisiana Purchase from France in 1803, which added vast western territories and more people to the United States, did not increase opportunities for blacks in the military. In fact, it diminished existing opportunities for some. Within the new territory lay New Orleans, with its large population of free blacks, many of whom had served in militia companies which had fought against both the British and Indians.

Blacks had served in the Louisiana militia while Spain claimed the territory and had fought under the command of Spanish governor Bernardo de Galvez in support of the American Revolution. Blacks continued to serve in local militias after France's takeover of the region as well. When the United States took possession of the area, the military and the new territorial government initially ignored the black militiamen until they could reorganize them. When efforts at restructuring the militia units failed, the government disbanded them altogether. Within a short time, the statutes of the 1792 Militia Act, which permitted only whites to serve, became the local policy.

Despite overwhelming obstacles, African Americans nevertheless contributed significantly to the American expansion toward the Pacific. York, a slave belonging to Capt. William Clark, accompanied the Lewis and Clark expedition to the Northwest in 1804–5. York impressed many of the Indian tribes, who believed the color black symbolized bravery, and assisted in the expedition's safe passage westward. Other blacks, including Edward Rose, Pierre Bonga, and James Beckwourth, ventured to the West as trappers and traders and later served as civilian scouts and guides for the military.

In the years following the turn of the century, the military continued to reject black Americans. Only when war loomed again in 1812 were African Americans asked to fill the ranks.

Great Britain continued to claim the right to stop and search U.S.-flagged merchant vessels for British naval deserters. Besieged by personnel shortages because of their involvement in the Napoleonic Wars, the British often impressed sailors from American ships regardless of nationality.

Just as Crispus Attucks had shed the first blood of the Revolution, African Americans were involved in the first major incident leading to the War of 1812. On June 22, 1807, the American frigate *Chesapeake* set sail after a refit at the Norfolk, Virginia, navy yard. Just offshore, the HMS *Leopard* fired a gun volley to stop the ship and then boarded to search for deserters. The boarders removed four sailors, allegedly former crewmen of the British ship *Melampus*, to the *Leopard*. Of the four, one was a British subject; the remaining three were African Americans.

The *Chesapeake* affair greatly angered Americans because they resented Britain's failure to honor U.S. maritime rights and the impressment of its residents. Also lurking in the minds of some Americans was the desire for additional expansion into Florida, the Northwest, and Canada. When the British continued to harass American shipping and impress its sailors, the United States declared war on June 12, 1812.

Hoping to benefit from Britain's preoccupation with its conflict on the Continent, Americans were nevertheless ill prepared for war. As a result, the American army initially experienced failure in its land battles. Although congressional acts of 1811, 1812, and 1814 called for "able-bodied, effective men," black volunteers found the ranks closed to them, since the army, despite its defeats, restricted enlistment to whites.

Some white leaders disagreed with the practice and lobbied for the enlistment of African Americans. One officer, Alexander Bill of Vermont, wrote to Secretary of War John Armstrong to explain his support of the right of blacks to serve and fight: "The Negroes of New England are actuated by principles equally honorable with those of the whites, and differ from them in nothing but the tincture of their skin." Bill's request went unheeded, as did similar ones from field commanders.

Thus, blacks rarely participated in the war during its first year, when most of the land battles took place along the Canadian border. In 1814, however, whites reversed their attitude when British forces attacked and burned the Capitol at Washington, D.C. Many white citizens in New York and Pennsylvania, who had formerly opposed blacks in the military, now welcomed them into the ranks to defend their cities against the advancing enemy.

Once again need overcame prejudices. The New York State government, on October 24, 1814, passed measures to raise "two regiments of color," comprised of two thousand men, with white officers in command. Freemen could enlist as well as slaves if their owners, who would receive their enlistment bonuses and wages, permitted. After three years of service, slaves would be given their freedom.

The city of Philadelphia also authorized the formation of a black battalion to bolster its defenses against the approaching British army. Neither this battalion nor the New York black regiments, however, ever saw combat. Before the military could form and train the black units, the British agreed to a peace treaty that ended the war.

While black soldiers played only a minor role in land combat, African Americans participated in every major naval battle of the War of 1812. Although officially excluded from the navy since the congressional act of 1798, blacks were able to join because difficulties in recruiting white crewmen encouraged captains to enroll sailors regardless of color. Finally, the continued manpower shortages forced the navy to officially authorize the recruitment of free blacks on March 3, 1813.

Blacks flocked to the navy and within months represented at least 10 percent, and possibly 20 percent, of American sailors. Many commanders welcomed the African-American sailors. Comdr. Thomas McDonough credited much of his success against the British on Lake Champlain to the accuracy of his gunners—mostly black volunteers.

Nathaniel Shafer, captain of the privateer *Governor Tompkins*, found his black crewmen exceptionally brave. After a devastating battle in the Atlantic Ocean, Shafer wrote: "The name of one of my poor fellows who was killed ought to be registered in the book of fame, and remembered with reverence as long as bravery is considered a virtue; he was a black man by the name of John Johnson; a 24-pound shot struck him in the hip and took away the lower part of his body; in this state the poor brave fellow lay on the deck and several times exclaimed to his shipmates, 'Fire away my boys, do not haul a color down.' "

Shafer also noted that another wounded black sailor asked to be thrown overboard because he felt "he was only in the way of others." The white ship captain concluded his report: "While America has such tars [sailors], she has little to fear from tyrants of the ocean."

Not all American naval commanders shared such respect for their black sailors—at least not initially. Oliver Hazard Perry, in command of a flotilla on the Great Lakes, stated that he welcomed anything "in the shape of a man" but nevertheless described arriving replacements as "a motley crew of blacks, soldiers, and boys."

Perry's commander, Comdr. Isaac Chauncey, did not take kindly to his subordinate's comments, informing Perry that he had fifty black crew members aboard his own ship "and that many of them are among my best men."

The commodore concluded with an observation that many American military leaders would not form for years to come when he said, "I have yet to learn that the color of a man's skin or the cut and trimmings of the coat can affect a man's qualifications or usefulness."

On September 10, 1813, Perry engaged the British fleet near Put-in-Bay, an island in Lake Erie off Ohio; after a savage three-hour battle he was ultimately victorious. One-fourth of Perry's four hundred crewmen were black. Best remembered for his victory message "We have met the enemy, and they are ours," Perry also included praise for his black sailors, acknowledging their contributions in the fight by stating in his battle report that his black seamen "seemed to be absolutely insensible to danger" and specifically mentioned African Americans Cyrus Tiffany, Jessie Walle, and Abraham Chase for their individual bravery.

The American naval victories, in which black sailors played such a critical role, finally forced the war-weary British to agree to a peace treaty in late 1814. However, because the only means of communication were by land or ship-borne messenger, the news of the peace was slow to reach the U.S. garrisons on its far borders. As a result, the final battle of the War of 1812 and the most significant land contribution by blacks to the conflict occurred at New Orleans after the war formally concluded.

Louisiana had become a state just months before the war and had disbanded its black militia units. When the war started, the British began actively recruiting former black militiamen. In response to the growing threat presented by the British regulars and their recruitment of local blacks into their ranks, Louisiana governor William C. Claiborne authorized the formation of four companies of sixty-four men each, the Battalion of Free Men of Color. Each black recruit had to be a free man who had paid taxes for at least two years and owned property valued in excess of three hundred dollars. The unit's authorization called for white officers in command, but within six months Claiborne appointed Isidore Honore, described as "a free man of color," second lieutenant. Claiborne shortly commissioned two more black men to join Honore as the first African American officers in any U.S. state militia.

Since no British attack against Louisiana materialized in 1812 or in the following year, the Battalion of Free Men of Color, restricted by law to their own state's borders, did not see action. In the fall of 1814, however, New Orleans appeared to be the next British target. Governor Claiborne contacted Gen. Andrew Jackson, the U.S. Army commander of the region, and offered the black battalion to support federal troops.

Whites across the South protested Claiborne's proposal, but Jackson welcomed the additional soldiers and began to recruit even more black soldiers. Jackson's primary motivation was twofold: He needed to increase his own numbers while denying the enemy potential recruits and also sought to counter British efforts to create unrest among the local black population. His address "to the Free Colored Inhabitants of Louisiana," dated September 21, 1814, and published in October, provided one of the most eloquent and poignant appeals to free African Americans to serve in the military. Jackson wrote:

> Through a mistake in policy, you have heretofore been deprived of participation in the glorious struggle for national rights in which our country is engaged. This no longer shall exist. As sons of freedom, you are now called upon to defend our most inestimable blessing. As Americans, your country looks with confidence to her adopted children for a valorous support, as a faithful return for the advantages enjoyed under her mild and equitable government. As fathers, husbands, and brothers, you are summoned to rally around the standard of the Eagle, to defend all which is dear in existence. Your country, although calling for your exertions, does not wish you to engage in her cause without amply remunerating you for the services rendered. Your intelligent minds are not to be led away by false representations. Your love of honor would cause you to despise the man who would attempt to deceive you. In the sincerity of a soldier and the language of the truth, I address you. To every noble-hearted, generous freeman of color volunteering to serve during the present contest with Great Britain, and no longer, will be paid the same bounty, in money and lands, now received by the white soldiers of the United States, viz., $124.00 in money, and 160 acres of land. Due regard will be paid to the feelings of freemen and soldiers—and will not, by being associated with white men in the same corps, be exposed to improper comparisons or unjust sarcasm. . . .

Jackson made no mention of African Americans still in bondage. Nevertheless, hundreds of free blacks came forward to join his army.

By the time the Americans met the advancing British at Chalmette Plains on January 8, 1815, more than six hundred black soldiers constituted about 10 percent of Jackson's total force. Through advantageous use of the terrain and defenses consisting of mounds of earth and cotton bales, the Americans dealt the British their worst defeat of the war, inflicting more than fifteen hundred casualties, compared with less than sixty of their own.

Despite the futility of the fight—the December peace treaty should have prevented the battle from occurring—Jackson kept his promises of money and land to his soldiers, both white and black. Many of the black soldiers also treasured copies of an address Jackson delivered to them at a review on December 18, 1814:

> To the men of color—Soldiers! From the shores of Mobile I collected you to arms. I invited you to share in the perils and to divide the glory of your white countrymen. I expected much from you, for I was not uninformed of those qualities which much render you so formidable to an invading foe. I knew you could endure hunger and thirst and all the hardships of war. I knew that you loved the land of your nativity, and that, like ourselves, you had to defend all that is most dear to man. But you surpass my hopes. I have found in you, united to these qualities, that noble enthusiasm which impels to great deeds. Soldiers! The President of the United States shall be informed of your conduct on the present occasion; and the voice of the representatives of the American nation shall applaud your valor, as your General now praises your ardor.

The men of the Louisiana Battalion of Free Men of Color and other African-American volunteers enjoyed a brief moment as heroes until word reached New Orleans that the war was over. Their burst of glory faded quickly, and once more the whites mustered them out of uniform to face prewar prejudices.

One hundred years after the Battle of New Orleans, the city celebrated the centennial of the victory. White soldiers marched, white orators spoke, white schoolchildren sang, and even white English visitors, representing the former enemy, occupied seats of honor. Not a single black person participated in the festivities, nor was any mention made of the Battalion of Free Men of Color or of any other African-American actions in the battle.

The lives of African Americans in Louisiana as well as those across the United States did not improve. White America quickly forgot or ignored the contributions of black soldiers. Although a congressional act of March 3, 1815, authorized a postwar army of ten thousand "able-bodied men," African Americans were not among those recruited to enlist. A War Department memorandum, published in the same year, expressed the general feelings of the white population by quoting a Boston register of discharges, who said, "A Negro is deemed unfit to associate with the American soldier."

In 1820 the U.S. Army issued a more specific order: "No Negro or mulatto will be received as a recruit of the Army." A year later, the Army General Regulations of 1821 limited entry into the service to "all free white male persons." State militias either officially or unofficially adopted the same policy.

While excluded from active service in the army, some African Americans did influence postwar military operations, not as a part of the army but as a major factor in the Seminole Wars. Florida, a Spanish territory, had become a haven for escaped slaves during and after the War of 1812. Supported first by the British and then by the Spanish, runaway slaves found refuge with Florida's Seminole Indians, where many intermarried and assumed leadership positions in the tribe.

In 1813 the Indians and runaway slaves occupied an abandoned British outpost, which became known as Negro Fort, just sixty miles from the U.S. border. Its proximity to the Deep South drew escaped slaves from throughout the region. Initiating what would become the First Seminole War, Andrew Jackson, the hero of the Battle of New Orleans, directed Col. Edmund P. Gaines to attack the fort in 1816 and "return the stolen Negroes and property to their rightful owners."

Gaines's cannon shots exploded the fort's ammunition magazine and killed most of the three hundred defenders. Two years later, Jackson personally lead another expedition into Florida, recapturing slaves and destroying Indian villages, thus bringing the First Seminole War to an end.

On February 22, 1819, the United States purchased Florida from Spain, a pivotal event for blacks in Florida, as well as their Indian allies. In response to the 1835 U.S. expeditions to "pacify" the Indians and reclaim slaves, African Americans in Florida staunchly fought back and held the army at bay for seven years. Many of the American commanders noted in their reports that the former slaves were the most skillful and resolute of the enemy forces, undoubtedly because the blacks were fighting for their continued freedom and surrender meant a return to bondage. But ultimately the blacks and Indians could not stop the U.S. Army, though their defeat did not come cheaply. Subjugation of blacks and Indians in Florida cost the United States $40 million and eighteen hundred lives.

When the Second Seminole War finally concluded in 1842, the army transplanted many of the Seminoles, including some of mixed Indian and black blood, to reservations in Oklahoma Territory. A few escaped into the deep Florida swamps to avoid capture and established a small Seminole community that survives today.

Soldiers who faced the Indians and black warriors learned that African Americans could indeed fight and fight well. Concerning such fighting abilities, veteran general Rufus Saxton wrote to a fellow officer: "I never had any doubts, Colonel, since my services in the Seminole Wars. There were many fugitive slaves among the Indians who fought us in the Everglades; in fact, I realized that was why we had been ordered to attack. The Negroes would stand and fight back, even with bare hands."

Even when directly confronted with the evidence, most whites were unwilling to concede that blacks were able to fight. One instance of such denial occurred after a small band of Indians and African Americans escaped to Mexico when the Seminoles were moved to Oklahoma. These renegades later made a few raids across the border into Texas, where they bested the locals in several fights. The Texans, so skeptical that blacks could fight that successfully—and so embarrassed by their inability to defeat the transplanted Floridians—rationalized that their elusive opponents must have been dark-skinned refugee Marmeduke soldiers from the Ottoman Empire.

Black Americans also continued to serve in the navy both during and after the War of 1812. At the end of the war, 10 percent of the U.S. Navy's personnel were black. The hardships of life at sea created a degree of equality among shipmates, which encouraged blacks to continue to enlist. The numbers of black sailors certainly must have included runaway slaves, but in an 1816 act, the navy officially prohibited slaves from serving on ships or working in shipyards. Apparently this edict did not prevent some slaves from being loaned to the navy, particularly in port jobs, by their owners, who then claimed their wages.

Usher Parsons, a veteran navy surgeon, recorded in his memoirs: "In 1816, I was surgeon of the *Java* under Commodore Perry. The white and colored seamen messed together. About one in six or eight were colored. In 1819 I was surgeon of the *Guerriere*, under Commodore Macdonough; and the proportion of blacks was about the same in her crew. There seemed to be an entire absence of prejudice against the blacks as messmates among the crew. What I have said applies to the crews of other ships that sailed in squadrons."

As more blacks joined the navy, complaints from white civilians and officers, based on the loss of jobs and simple racism, flooded Congress and the Department of the Navy. In response, the navy issued an order in 1839 restricting black enlistments to no more than 5 percent of the total. Ironically, Isaac Chauncey, who had commended the performance of his black sailors to Oliver Perry during the War of 1812, signed the order as the acting secretary of the navy. Two years later, Secretary of the Navy Abel P. Upshur reported to Congress that the number of black sailors on active duty remained well within the 5 percent quota.

Neither the official exclusion of slaves nor the 5 percent enlistment quota of free blacks placated all white Americans, particularly those in the South. In 1842, South Carolina senator John C. Calhoun introduced legislation restricting blacks, regardless of status, to positions of cooks, mess boys, and servants. The honorable senator expressed the feelings of many of his fellow Americans in both the South and the North when he stated that "those who have to sustain the honor and glory of the country" should not be "degraded by being mingled and mixed up with an inferior race." The measure passed the Senate, but the northern-dominated House of Representatives never brought the bill to a vote.

Although their numbers remained small, black sailors continued to share the dangers and hardships of sea duty with their white shipmates. On November 24, 1846, Capt. J. H. Aulick, of the *Potomac*, reported to his commanding officer: "Sir, I regret to inform you of the death of two of my crew—Sam Thomas, seaman (colored), died on the 16th of dysentery. It is said he was a native of New London at which place he has a family. The other, John Crook, ordinary seaman (colored) died this morning from the effect of injuries received by a fall. It is believed he was a citizen of Baltimore at which place he has relatives."

Unlike the navy, the U.S. Army remained "all white" after the Seminole Wars. Some 160,000 Americans served in the Mexican War of 1846–48; 13,000 died in battle or from disease. Officially, none were black, because the military did not allow African Americans to serve. Most accounts of the war state, if they make any mention of the fact, that no blacks served in the war. Actually, several did join the American forces in Mexico as servants to white officers or in other support roles. History, however, has failed to record their names or any specific accomplishments.

Although relegated to only a footnote in the official naval histories of the Mexican War, at least a thousand black seamen also served during the conflict aboard American warships blockading Mexican ports. They served, too, as crew members of vessels delivering men and supplies to U.S. forces on the east coast of Mexico.

However, by the middle of the nineteenth century, white Americans had largely forgotten the past performance of blacks in the military and for the most part ignored the ongoing service of African-American sailors. Black abolitionist and historian William C. Nell published *The Colored Portrait of the American Revolution* in 1855 detailing the contributions of African Americans in the Revolutionary War, the War of 1812, and the other conflicts of the period. In his preface Nell wrote, "A combination of circumstances have veiled from the public eye a narration of those military services which are generally conceded as passports to the honorable and lasting notice of Americans."

While their contributions in the War of 1812 and the Mexican War faded into oblivion, African Americans were soon to play an important role in their country's most divisive conflict. Even before the Revolutionary War, the North and the South were divided by geographic and political differences. America was on a collision course with itself.

3

THE AMERICAN CIVIL WAR

For the third time in as many wars fought on U.S. soil, the first casualty of the Civil War was an African American. He died in October 1859 when John Brown led a raid on Harpers Ferry, West Virginia, to secure weapons to arm a slave rebellion. During the initial assault on the town's federal arsenal, Brown's raiders accidentally killed Hayward Shepherd, a free black railway baggage master. Although the raid failed and Brown died on the gallows, the incident focused the country's attention on the divisive issue of slavery and the fate of African Americans.

For white Southerners, slavery was an economic issue. The economy of the region was dependent on the production of cotton and other crops on large plantations, which were directly dependent on slave labor for their profits. Abolishing slavery would end the plantation system and thus the economic base of the South. For many white Northerners, however, slavery was a human-rights issue. The diversified agricultural-industrial base of that region was dependent on hired labor. Abolishing slavery would have less impact on economics in the North.

The political and cultural differences between North and South that John Brown had attempted to exploit came to a head with the election of Abraham Lincoln in 1860. Although neither Lincoln nor his Republicans championed the rights of African Americans, they were the more sympathetic party, and their victory led South Carolina and six other states to secede from the Union and form the Confederate States of America even before the new president could assume office on March 3, 1861.

As more states left the Union and joined the Confederacy, few Americans of either side thought that their differences would lead to war, nor did either side directly confront the issue of slavery. Southerners claimed that seceding and forming their own government was well within their authority under states' rights. Northerners, indicating that their objective was to preserve the Union, presumed that a naval blockade and diplomacy would soon return the wayward Southern states to the federal fold. Even the bombardment and fall of Fort Sumter, South Carolina, on April 14, 1861, did not alert white Americans that a long and bloody war lay ahead.

From the time the first twenty slaves, in shackles, took up their residence in the Virginia colony in 1619, the question of a man's right to "own" another became interwoven into the fabric of American culture. Because generation after generation of whites failed to reach a unanimous agreement on slavery, individuals and groups devised their own justifications, basing their rationales more often on economic and social issues than moral concerns. By the nineteenth century, these "justifications" had solidified into beliefs and prejudices which, once established, were difficult, if not impossible, to change.

The majority of the black population, 3,500,000, lived in the South as slaves. Southern whites rationalized that slavery benefited blacks as well as whites, which, in turn, justified whatever restrictions they placed on their chattel. After more than two hundred years, for many white Americans slavery was simply a way of life inherited from their ancestors, and most saw no reason to change it.

To ensure the continuance of the status quo, states throughout the South passed laws prohibiting the teaching of a slave to read and write. A typical justification of this restriction appeared in the 1831 statute passed by the General Assembly of the State of North Carolina, which began, "Whereas the teaching of slaves to read and write, has a tendency to excite dissatisfaction in their minds, and to produce insurrection and rebellion, to the manifest injury of the citizens of this State. . . ."

White Southerners who might have questioned "the virtues" of slavery found abundant "authority" to quell those doubts. Virginia lawyer

George Fitzhugh wrote several books and articles in the 1850s expounding on what he called "the universal law of slavery." Fitzhugh expressed a belief in the natural inequality of men and stated that both blacks and whites benefited from slavery. Fitzhugh declared:

> He the Negro is but a grown up child, and must be governed as a child, not as a lunatic or criminal. The master occupies toward him the place of parent or guardian. . . . The Negro is improvident; will not lay up in summer for the wants of winter; will not accumulate in youth for the exigencies of age. He would become an insufferable burden to society. Society has the right to prevent this, and can only do so by subjecting him to domestic slavery. In the last place, the Negro race is inferior to the white race, and living in their midst, they would be far outstripped or outwitted in the chaos of free competition.

John H. Van Evrie, a prominent Washington, D.C., physician, supported Fitzhugh's "universal law" with his "plurality" theory of the origins of humans, published in 1853. Van Evrie claimed that the brain of a black person was significantly smaller than that of a white person, which explained why the black race had "never of its own volition passed beyond the hunter condition." The doctor elaborated:

> The Negro is a man, but a different and inferior species of man, who could no more originate from the same source as ourselves, than the owl could from the eagle, or the shad from the salmon, or the cat from the tiger; and who can no more be forced by human power to manifest the faculties, or perform the purposes assigned by the Almighty Creator to the Caucasian man, than can either of these forms of life be made to manifest faculties other than those inherent, specific, and eternally impressed upon their organization.

On March 4, 1858, John Henry Hammond, a wealthy South Carolina plantation owner, delivered a speech to the U.S. Congress justifying slavery and claiming that bondage actually improved the condition of blacks over their brothers in Africa. Hammond further described slaves in the South as "happy, content, unaspiring, and utterly incapable, from intellectual weakness" of giving their masters any trouble. He then explained his "mud-sill" theory that any viable society must have two groups, one superior and one inferior. Hammond wrote:

> In all social systems, there must be a class to do the menial duties, to perform the drudgery of life. That is, a class requiring but a low order of intellect and but little skill. Its requisites are vigor, docility, fidelity. Such a class you must have, or you would not have the other class which leads progress, civilization, and refinement. It constitutes the very mud-sill of society and of political government; and you might as well attempt to build a house in the air, as to build either the one or the other, except on this mud-sill. Fortunately for the South, she found a race adapted to that purpose to her hand. A race inferior to her own, but eminently qualified in temper, in vigor, in docility, in capacity to stand the climate, to answer all her purposes.

Southerners believed that Lincoln's election threatened slavery. In a March 12, 1861, speech in Savannah, Georgia, Confederate vice president Alexander Stephens proclaimed that slavery was the "immediate cause" of Southern secession, stating, "Our Confederacy is founded upon . . . the great truth that the Negro is not equal to the white man. That slavery—subordination to the superior race, is the natural and normal condition. This, our new Government, is the first, in the history of the world, based upon this great physical and moral truth."

Lincoln, on the other hand, made no official mention of slavery as the pivotal issue as he armed the North following Fort Sumter. Despite this omission, African Americans came forward in droves to volunteer to fight to preserve the Union. On April 17 free blacks in Pittsburgh formed what they called the Hannibal Guards and volunteered their service to Gen. James S. Negley, the commander of the Western Pennsylvania militia. Their letter to Negley provides an excellent summary of the feeling of free Northern blacks of the period:

> Sir: As we sympathize with our white fellow-citizens at the present crisis, and to show that we can and do feel interested in the present state of affairs; and as we consider ourselves American citizens and interested in the Commonwealth of all our white fellow-citizens, although deprived of our political rights, we yet wish the government of the United States to be sustained against the tyranny of slavery, and are willing to assist in any honorable way or manner to sustain the present administration. We therefore tender to the state the services of the Hannibal Guards.

Only a week after the fall of Fort Sumter, Washington, D.C., freeman Jacob Dobson sent a letter to the secretary of war declaring: "I desire to inform you that I know of some three hundred of reliable colored free citizens of this city, who desire to enter the service for the defense of the city. I have been three times across the Rocky Mountains in the service of the country with Frémont and others. I can be found about the Senate Chambers, as I have been employed about the premises for some years."

At about the same time, freemen in Cleveland met to declare their patriotism and willingness to serve: "Resolved, that we as colored citizens of Cleveland, desiring to prove our loyalty to our government, feel that we should adopt measures to put ourselves in a position to defend the government of which we claim protection. . . . As in the times of '76 and the days of 1812, we are ready to go forth and do battle in the common cause of the country."

In Boston blacks petitioned the state legislature for the right to serve as they had in previous conflicts. Black residents of New York formed their own unit and began drilling in anticipation of being called to active duty. In Cincinnati blacks organized a company of Home Guards and volunteered their service for the defense of their city.

Once again African Americans were ready and willing to serve in the military where needed. This time, the war was *about*, and *over*, them. Unfortunately, that was the only difference, because once more, as in the past, blacks encountered opposition to their service.

White leaders recognized none of the black volunteer efforts and enlisted not a single African-American soldier. The Cincinnati police chief made a typical response, telling the Home Guards, "We want you damn niggers to keep out of this; this is a white man's war."

Federal officials expressed similar views. In his response to the previously mentioned request of Senate employee and western exploration veteran Jacob Dobson, Secretary of War Simon Cameron responded, "This Department has no intention at present to call into service of the Government any colored soldiers."

President Lincoln supported the exclusion of black volunteers and shared the belief of many of his subordinates that the war would be of short duration. In fact, his first call for seventy-five thousand volunteers required an enlistment period of only ninety days. Lincoln not only felt it was unnecessary to recruit blacks; he also feared that to do so might push the slaveholding border states into the Confederate fold. Gov. David Tod of Ohio voiced still another consideration behind the president's decision.

The governor asked, "Do you know that this is a white man's government; that the white men are able to defend and protect it; and that to enlist a Negro soldier would be to drive every white man out of the service?"

During the first few months after the fall of Fort Sumter, President Lincoln forbade the enlistment of black soldiers; he and his military commanders made no mention of abolishing slavery. In the border state of Maryland, Gen. Benjamin F. Butler offered federal troops to the governor to suppress a possible slave rebellion. When Union troops landed in South Carolina, their commander, Gen. Thomas W. Sherman, issued a proclamation promising not to interfere with "social and local institutions"—in other words, slavery.

Shortly before the end of the ninety-day enlistments in Lincoln's volunteer white army, Confederate forces decisively defeated a Union force at the Battle of Bull Run, Virginia, on July 21, 1861. Military personnel and government leaders on both sides began to realize that what they had considered a dispute was, in fact, developing into a full-fledged war.

Still Lincoln did not include abolition in his war objectives. Nevertheless, civilian abolitionists continued their efforts to influence emancipation and to encourage enlistment of black servicemen. The most eloquent black freeman, Frederick Douglass, noted: "Colored men were good enough to fight under Washington, but they are not good enough to fight under McClellan." Douglass also addressed the possibility of the South arming its own African Americans to help preserve the Confederacy, prophesying, "The side which first summons the Negro to its aid will conquer."

Even though Union leaders in Washington officially ignored Douglass, several generals in the field began efforts to free slaves and welcome blacks into their ranks. In the border state of Missouri, Union commander and avid abolitionist John C. Frémont went as far as to declare all blacks free and to send his wife to Washington to lobby for the president's support in freeing Missouri's slaves. Southern sympathizers in the state organized a guerrilla campaign against Frémont, so that in the end all his declaration accomplished was to add to intraregional hostilities.

Company E, Fourth U.S. Colored Infantry Regiment, Fort Lincoln, Washington, D.C. (ca. 1864). (LIBRARY OF CONGRESS) ***Picture framed under broken glass plate.***

Meanwhile, repeated Confederate victories and mounting Union casualties were moving more Northerners to support the military induction of African Americans. As early as the fall of 1861, Gov. John A. Andrew of Massachusetts summed up the growing feeling in the North when he told a New York audience: "It is not my opinion that our generals, when any man comes to the standard and desires to defend the flag, will find it important to light a candle, and see what his complexion is, or to consult the family Bible to ascertain whether his grandfather came from the banks of the Thames or the banks of the Senegal."

By the middle of 1862 newspapers across the North had joined the abolition movement and were supporting the rights of blacks to enlist in the military. Whereas some believed in equality, the *Philadelphia North American* openly declared, "The lives of white men can and ought to be spared by the employment of blacks as soldiers."

Despite pressure from politicians and newspapers, Lincoln held firm in his exclusion of blacks from the military when he issued a call for 300,000 volunteers in July 1862. A few weeks later, after Congress revoked state militia laws excluding black enlistments and approved their use as laborers and in other support positions, Lincoln concurred with the legislation but insisted that blacks remain restricted from combat roles. Still concerned about more states leaving the Union for the Confederacy, the president explained that arming blacks "would turn fifty thousand bayonets from the loyal border states against us."

Despite Lincoln's reluctance to enlist African Americans, General Butler, who had recently offered to quell slave rebellions, now encouraged slaves to cross over Union lines to work as laborers in building fortifications. Butler called these runaways "contrabands of war," a label which stuck.

On the Sea Islands off the Georgia coast, Gen. David C. Hunter organized "contrabands" into military units as early as May 1862. Hunter's recruitment practices, more akin to impressment, resulted in a reluctance by blacks to make themselves available for his forced service. Nevertheless, Hunter managed to field a company-sized unit of former slaves to perform garrison support duty on Saint Simons Island. Hunter provided few arms to this company of contrabands.

It was not until the easing of restrictions on black recruiting the following July that Gen. Rufus Saxton, replacing Hunter, began to expand the number of local Sea Island blacks and arriving runaways and to arm them appropriately. By late August 1862, Saxton had War Department permission to recruit, arm, and equip a regiment of five thousand African Americans, led by white officers. He declared each volunteer, along with his wife and children, "forever free" in recognition of his enlistment. Saxton established the headquarters of the First South Carolina Colored Volunteers at Port Royal, South Carolina, and sent for Thomas Wentworth Higginson, a Boston abolitionist and friend of the late John Brown, to command the unit. Officially, the United States did not call the First South Carolina Colored Volunteers into active duty until January 31, 1863, making them the fifth black regiment so recognized. In reality, they have an honest claim to being the first black combat unit organized in the war, regardless of official dates.

Union commanders in the West also pushed to form black fighting units. Kansas Unionist James H. Lane, despite opposition from the War Department, began recruiting a black regiment to fight Confederate guerrilla William Quantrill in August 1862. Lane directed his recruiters to enlist those blacks "evincing by their actions a willing readiness to link their fate and share the perils with their white brethren in the war of the great rebellion."

Lane's recruitment went well. On August 17 a correspondent for the *New York Times* reported, "Under his [Lane's] direction [blacks] are enthusiastic to enlist and fight." Within weeks Lane had five hundred volunteers formed into the first black regiment recruited and organized in a free state. About half of the regiment engaged in the first significant skirmish between Union blacks and Confederate rebels on October 28, 1862, at Island Mound near Butler, Missouri. On January 13, 1863, the First Kansas Colored Volunteers were officially mustered into federal service.

Following the traditions of free blacks participating in the defense of New Orleans that predated the War of 1812, African Americans formed regiments in that city even earlier than those organized in South Carolina and Kansas. When Louisiana seceded from the Union, that state's governor, Thomas O. Moore, authorized a regiment of free blacks, called the Native Guards, to defend New Orleans against any Union invasion. Jordan Noble, once a drummer boy in the 1815 Battle of New Orleans, helped raise one of the companies.

When Union admiral David Glasgow Farragut forced his way up the Mississippi River from the Gulf of Mexico and his capture of New Orleans became inevitable in April 1862, the Native Guards refused to evacuate the city to join Confederate defenses elsewhere in the state. The black militiamen stated that their responsibility was the defense of their city, not Louisiana or the Confederacy. After the Union forces succeeded in taking the city, the Native Guards sided with the victors and joined the ranks of Gen. Benjamin Butler, who was in yet another occupation assignment.

Butler, mostly acting on his own and often in conflict with other Union commanders in the area, issued General Order No. 63, on August 22, calling on additional freemen in Louisiana to join the Native Guards. Both to appease his fellow officers and to avoid enraging the strong Confederate forces still operating nearby, Butler limited his call for volunteers to free blacks, making no mention of those still held in, or recently escaped from, slavery. On September 1, Butler informed the War Department, "I shall also have within ten days a regiment of 1,000 strong of Native Guards (Colored). . . ." On September 27 the First Regiment of the Louisiana Native Guards was mustered into federal service, followed by the Second Regiment on October 12 and the Third Regiment on November 24, making them the first official three black regiments in the Union army.

Despite the successes of these black regiments, Lincoln remained reluctant to support the recruitment of African Americans into the military or even to include the ending of slavery as an objective of the conflict. By the summer of 1862, however, Lincoln was feeling increasing pressure from the press—particularly Horace Greeley of the *New York Tribune*—to take a strong stand on emancipation. In response, Lincoln clarified his position, stating, "My paramount object in this struggle is to save the Union, and is not either to save or to destroy slavery. If I could save the Union without freeing any slave I would do it; and if I could save it by freeing all the slaves I would do it; and if I could save it by freeing some and leaving others alone I would also do that."

Even as he made that statement, the war itself was pushing the president into taking a decisive position. After more than a year of continuous victories by the Confederates, the demoralized Union army was experiencing difficulties in attracting volunteers to fill its rapidly depleting ranks. Lincoln recognized that freeing the slaves was becoming essential to his objective of preserving the Union.

The president, however, kept his thoughts to himself as he began drafting a proclamation of emancipation during June 1862. On July 22, Lincoln met with his cabinet and, without preamble, read the proclamation. He was well into the document before his cabinet officers realized its significance. Some favored the proposal; others opposed it; all feared its political ramifications in the fall elections.

The cabinet debated the issue for several hours, and the president met with several of his staff over the next few days to continue the conversations. Finally, Lincoln made up his mind to proceed with the Emancipation Proclamation but decided, "We mustn't issue it till after a victory."

While the president and his cabinet wrestled with their decision, the rebels continued to dominate the battlefield. It was not until September 17, 1862, when the Union army turned back Gen. Robert E. Lee's invasion of the North at Sharpsburg, Maryland, that President Lincoln finally had his victory. On September 22 he announced his plan to issue the Emancipation Proclamation, to take effect on January 1, 1863.

Interestingly, the Emancipation Proclamation had little direct effect on the lives of most African Americans at the time. The proclamation, more a political tactic than a step toward equality, freed slaves only in Rebel-held territory—the very places where neither Lincoln nor his army had authority or power to enforce the act. Provisions of the proclamation did not apply to slaves held in non-Confederate territory. Even though the vast majority of America's slaves would not be free until the Thirteenth Amendment passed two years later, African Americans celebrated the Emancipation Proclamation for its intent and still revere it as their document of freedom. What the Emancipation Proclamation did do, however, was make every new Union offensive into the South a campaign of liberation; it also authorized increased induction of blacks into the federal forces.

The announcement of the Emancipation Proclamation produced celebration by African Americans across the country and spawned an immediate movement to actively enlist blacks into the armed forces. Massachusetts governor John A. Andrew, a zealous abolitionist, began lobbying the War Department for permission to organize black units in his state. On January 26, 1863, Andrew received authorization to form two regiments of black volunteers, to be led by white officers.

Despite the Emancipation Proclamation, both Robert Gould Shaw, commander of the Fifty-fourth Massachusetts, and Norwood P. Hallowell, commander of the Fifty-fifth Massachusetts, experienced early recruiting difficulties in meeting Governor Andrew's quota of volunteers. Few blacks resided in Massachusetts or the surrounding states, and many who did had acquired good jobs in the surging war economy. Unexciting recruiting posters proclaimed: Wanted. Good Men for the 54th Regiment of Massachusetts Volunteers of African Descent. Even promises of a hundred-dollar bonus, more than three months' wages for a civilian laborer, paid at expiration of the service term, brought few enlistees.

Shaw and Hallowell expanded their recruitment area to the entire United States and the occupied territories of the Confederacy, and they sought the assistance of abolition leaders, both black and white. Only then did large numbers of volunteers come forward. Frederick Douglass provided significant help in a March 2, 1863, address when he shouted, "Men of color, to arms," and argued that "liberty won only by white men would lose half its luster."

On May 28, 1863, the one-thousand-man Fifty-fourth Massachusetts Infantry Regiment, under Shaw's command, paraded through Boston before huge black and white crowds on their way to the front. According to the *Boston Evening Journal*, "no regiment on its departure has collected so many thousands as the Fifty-fourth. The early morning trains from all directions were filled to overflowing, extra cars were run, vast crowds lined the streets where the regiment was to pass, and the Common was crowded with an immense number of people, such as only the 4th of July or some rare event causes to assemble."

Governor Andrew, proud of his state's accomplishment, also realized the importance of the future performance of the Fifty-fourth. Shortly after the regiment's parade, he wrote, "Its success or failure will go far to elevate or depress the estimation in which the character of Colored Americans will be held throughout the war."

Hallowell's Fifty-fifth Massachusetts Infantry Regiment followed Shaw's to the front a few weeks later. Over the next months, black regiments from Rhode Island, Connecticut, Pennsylvania, Ohio, Illinois, Indiana, and Michigan also joined the Union army. All had white officers, many of whom received commissions from the enlisted ranks to assume junior-officer positions. Most of the senior officers advanced several grades as a reward for assuming black regimental leadership positions. These promotions included both Shaw and Hallowell, who rose from captains in white regiments to colonels in command of black units.

While the newly formed black units from the North moved south to join the war, members of the Corps d'Afrique, composed of the Louisiana Native Guards regiments that had been formed before the Emancipation Proclamation, engaged in their first major battle. On May 27, 1863, the First and Third Louisiana, supported by white engineer regiments, made multiple attempts to cross broken ground covered with downed trees and reach a Confederate artillery and infantry position at Port Hudson. Outnumbered and outgunned, the black soldiers failed to overrun the Rebel defenses, but their heroic actions and the loss of one-fifth of their numbers did not go unnoticed.

Gen. Daniel Ullmann, the white commander of the black Louisiana soldiers, reported that his men "made six or seven charges . . . against the enemy's works." He concluded: "They were exposed to a terrible fire and were dreadfully slaughtered. While it may be doubtful whether it was wise to so expose them, all who witnessed these charges agree that their conduct was such as would do honor to any soldiers. The conduct of these regiments on this occasion wrought a marvelous change in the opinion of many former sneerers."

White officers who participated in the battle echoed Ullmann's praise and gained newfound respect for the black soldiers. An engineer officer wrote: "You have no idea how my prejudices with regard to the Negro troops have been dispelled by the battle the other day. The brigade of Negroes behaved magnificently and fought splendidly; could not have done better. They are far superior in discipline to the white troops, and just as brave."

Senior Union commanders shared the opinions of the junior officers. Gen. Nathaniel P. Banks, not known as a friend of either Ullmann or of African Americans in general, praised the performance of the Louisiana regiments. Banks wrote his superiors:

> Whatever doubt may have existed heretofore as to the efficiency of Negro regiments, the history of this day proves conclusively to those who were in condition to observe the conduct of these regiments that the Government will find in this class of troops effective supporters and defenders. The severe test to which they were subjected, and the determined manner in which they encountered the enemy, leaves upon my mind no doubt of their ultimate success.

The *New York Times* and other Northern newspapers also cited the Battle of Port Hudson as proof of the fighting ability of black troops. The *Times*, which had withheld its support of the black regiments until they had been battle tested, quoted much of Banks's report to his superiors and commented editorially:

> This official testimony settles the question that the Negro race can fight with great prowess. Those black soldiers had never before been in any severe engagement. They were comparatively raw troops, and were yet subjected to the most awful ordeal that even veterans ever have to experience—the charging of fortifications through the crash of belching batteries. The men, white or black, who will not flinch from that, will flinch from nothing. It is no longer possible to doubt the bravery and steadiness of the colored race, when rightly led.

In their concluding words—"when rightly led"—the *Times* alluded to the need for white officers to lead black soldiers. Not many Americans were aware that the First and Third Louisiana were two of only a few regiments with black officers—a linkage to their original service as freemen in defense of New Orleans. Thirty-eight African American officers—fully two-thirds of the total number of company-grade officers engaged—saw action at Port Hudson.

Less than a month after Port Hudson, two other black Louisiana regiments won the Civil War's first significant battle secured by African Americans. On June 7, a Confederate force of two thousand to three thousand attacked the camp of the Ninth and Eleventh Infantry Regiments at Milliken's Bend, Louisiana. In the official account of the battle, Assistant Secretary of War Charles Dana reported, "The Negro troops at first gave way, but hearing that those of their number who were captured were killed, they rallied with great fury and routed the enemy."

Dana's report also included a quote from Gen. Elias S. Dennis, commander of the District of Northern Louisiana. About the defenders of Milliken's Bend, Dennis stated, "It is impossible for men to show greater gallantry than the Negro troops in this fight."

More revealing than the comments of their own commanders is the evaluation of the black soldiers by their Confederate enemies. The Rebel report of the fight stated that their own "charge was resisted by the Negro portion of the enemy's force with considerable obstinacy."

The Louisiana regiments fought the first major battles by African-American soldiers in the Civil War, but the Fifty-fourth Massachusetts gained the most widespread and lasting fame earned by a black unit during the conflict for its assault on Confederate Fort Wagner, South Carolina. Prior to the battle, the Fifty-fourth Regiment, conducting mostly support activities, had been involved only in minor skirmishes. Nevertheless, Gen. George C. Strong selected the Fifty-fourth to lead the attack. Conflicting reports credit different motivations for the Massachusetts regiment's selection as the advance unit.

In response to the question of how he planned to organize for the attack, Union commander Gen. Truman Seymour, replied, "Well, I guess we'll let Strong lead and put those damn niggers from Massachusetts in the advance; we might as well get rid of them one time or another."

Colonel Shaw had repeatedly requested more important duties to prove his regiment's worth, and Strong gave him the opportunity. What-

Rendition of the Fifty-fourth Massachusetts Infantry Regiment attacking Fort Wagner, South Carolina, on July 18, 1863. (U.S. ARMY MILITARY HISTORY INSTITUTE)

ever the reason, when offered the chance to lead the attack, Shaw willingly accepted and late in the afternoon of July 18, 1863, ordered his men forward. They marched toward the Confederate defenses along a narrow, three-quarter-mile-long road bordered on one side by the ocean and on the other by a deep marsh. With muskets, bayonets, and bare hands, the African Americans attacked, but the overwhelming numbers of the enemy did not yield. Within hours, Shaw and many of the men of the Fifty-fourth lay dead on the parapets of Fort Wagner. Follow-up attacks by the white regiments also failed to rout the Confederates. Among other casualties was General Strong.

The actions of twenty-three-year old sergeant William H. Carney of Company C typified the bravery exhibited by members of the Fifty-fourth Massachusetts at Fort Wagner. Early in the fight, when the regular regimental color-bearer became a casualty, Carney seized the flag, led the assault against the Rebel breastwork, and achieved the Union's deepest penetration of the Confederate lines. Carney, suffering two wounds before being driven back, stumbled to the rear, where a member of Strong's white brigade offered to help him carry the heavy standard. Carney refused, stating, "No one but a member of the Fifty-fourth should carry the colors."

When Carney returned the unit's flag to his cheering comrades, he proudly said, "The flag never touched the ground, boys." On June 30, 1864, Carney left the regiment and the army because of a lingering disability from his wounds. Before his discharge, the U.S. Army made him the first African-American recipient of the country's highest award for valor in combat, the Medal of Honor.

The victorious Confederates buried the dead of the Fifty-fourth Regiment in a common grave just outside their fort. Rumors swept across the Northern states that the Confederates had reported burying Shaw "with his niggers." Shaw's father, Francis, later stopped efforts to recover his son's body. He wrote: "We hold that a soldier's most appropriate burial place is on the field where he has fallen."

Although the assault on Fort Wagner had been a military failure, it provided still more proof that black soldiers could and would fight even in the most severe combat conditions. General Seymour explained that the choice of attacking with the Fifty-fourth "was in every respect as efficient as any other body of men, and as it was one of the strongest and best officered, there seemed to be no good reason why it should not be selected for the advance."

Northern politicians, newspapers, and military leaders also hailed the Fifty-fourth, making them the symbol of African-American participation in the war. In a letter to a fellow abolitionist, Angelina G. Weld wrote: "Do you not rejoice and exult in all that praise that is lavished upon our brave colored troops even by the pro-slavery papers? I have no tears to shed over their graves, because I see that their heroism is working a great change in public opinion, forcing men to see the sin and shame of enslaving such men."

Sgt. William H. Carney, Company C, Fifty-fourth Massachusetts Infantry Regiment: First African American awarded the Medal of Honor—for actions at Fort Wagner, South Carolina, on July 18, 1863. (U.S. ARMY MILITARY HISTORY INSTITUTE)

The performance of the Fifty-fourth Massachusetts at Fort Wagner also made some senior Union leaders aware that the number of able African Americans available to join the fight could significantly influence the war's outcome. On August 23, 1863, a little more than a month after the Battle of Fort Wagner, Gen. Ulysses S. Grant wrote in a letter to President Lincoln: "I have given the subject of arming the Negro my hearty support. This, with the emancipation of the Negro, is the heavyist [*sic*] blow yet given the Confederacy. . . . By arming the Negro we have added a powerful ally. They will make good soldiers and taking them from the enemy weakens him in the same proportion they strengthen us."

More than two years after the battle, the *New York Tribune* reflected on the importance of the Battle of Fort Wagner: "It is not too much to say that if this Massachusetts 54th had faltered when its trial came, two hundred thousand colored troops for whom it was a pioneer would never have been put in the field, or would not have been put in for another year, which would have been equivalent to protracting the war into 1866. But it did not falter. It made Fort Wagner such a name to the colored race as Bunker Hill has been for ninety years to the white Yankees."

More than a century later, Hollywood honored the Fifty-fourth's assault against Fort Wagner in a feature film titled *Glory*, starring Matthew Broderick, Morgan Freeman, and Denzel Washington. The movie, directed by Edward Zwick, earned an Oscar nomination for Best Picture in 1989.

Even before the Massachusetts and Louisiana black regiments proved themselves in battle, the United States had taken measures to recruit and organize additional black regiments. On May 22, 1863, the War Department authorized the Bureau of Colored Troops. By the time the Battle of Fort Wagner occurred, thirty black regiments were on active duty, and this number doubled by the end of the year, with the regiments mustered into federal service as a part of the U.S. Colored Troops (USCT).

Recruiting posters for the USCT called for black volunteers to come forward to free their brethren still held in captivity and to stand up for their race. Burlington County, New Jersey, mirrored other enlistment efforts with a poster directed to "colored men." It declared:

> Who would be free themselves must strike the first blow! Your county calls you to the Field of Martial Glory. Providence has offered you an opportunity to vindicate the patriotism and manhood of your race. Some of your brothers accepting this offer on many a well-fought field, have written their names on history's immortal page amongst the bravest of the brave. Now is your time! Remember, that every blow you strike at the call of your government against the accursed Slaveholder's Rebellion, you break the shackles from the limbs of your kindred and their wives and children.

During the battles of 1863, African-American soldiers established themselves as valuable members of the U.S. military. In a letter to Secretary of War Edwin M. Stanton in August, Judge Advocate General and former Secretary of War Joseph Holt summarized the accomplishments of the black soldiers: "The tenacious and brilliant valor displayed by troops of this race at Port Hudson, Milliken's Bend, and Fort Wagner have sufficiently demonstrated to the President and to the country the character of service of which they are capable. . . . In view of the loyalty of this race, and of the obstinate courage which they have shown themselves to possess, they certainly constitute, at this crisis in our history, a most powerful and reliable arm of public defense."

Despite their accomplishments on the field of battle and the accolades bestowed on them by the press, politicians, and military leaders, black soldiers still faced staunch opposition from many Northerners. Some opposed arming African Americans, others still doubted their ability to fight, and still others resented the "advancement" of the race, which meant competition for jobs and better wages.

Resentment against emerging rights for blacks exploded in the summer of 1863 in New York. Many of the city's citizens, particularly recent Irish immigrants, who disliked competing with blacks for employment, became further disgruntled as a result of the passage of the Conscription Act of March 1, aimed at drafting men into the Union army. One of the act's provision seemed particularly unfair to those in the lower economic classes: Anyone could buy an exemption for three hundred dollars, a sum far beyond the average worker's financial capabilities.

White citizens across the North began rioting in protest against conscription and the exemption clause. Looking for scapegoats to blame for the cause of the war and shrinking job opportunities, some of the rioters turned their anger against local African Americans. In New York City, during the same week that the Fifty-fourth Massachusetts assaulted Fort Wagner, whites rioted in the streets and directed their vengeance at the black community. They killed more than a hundred defenseless African Americans, including children in the city's Colored Orphan Asylum, which they burned to the ground. At least an additional one hundred black New Yorkers were injured, and thousands fled the city to the safety of the countryside.

The majority of white Northern soldiers and civilians, however, accepted African Americans as a part of the army. Some did so from a sense of morality, but far more advocated the "Sambo's Right to be Kilt" belief, which held that the black man had as much right to die as anyone else. The Sambo character with an "equal right to die" emerged in an early 1864 song credited to Pvt. Miles O'Reilly. In actuality, the author of the ditty was New Yorker Charles G. Halpine, a former officer in Gen. David Hunter's Department of the South. Halpine's verse declared:

Some tell us 'tis a burnin shame
  To make the naygers fight;
An' that the thrade of bein' kilt
  Belongs but to the white;
In battle's wild commotion
  I shouldn't at all object
If Sambo's body should stop a ball
  That was comin' for me direct;
Though Sambos' black as the ace of spades,
  His finger a thrigger can pull,
And his eyes runs straight on the barrel-sights
  From under its thatch of wool.
So hear me all, boys darlin',
  Don't think I'm tippin you chaff,
The right to be kilt we'll divide wid him,
  And give him the largest half!

While they had gained equality in facing death, blacks still found racism entrenched in other areas, including inequitable military pay. The source of the black soldiers' complaint was Section 15 of the Militia Act of July 17, 1862, which provided the only guideline for payment of soldiers. The act authorized a white private to receive thirteen dollars a month, with advances in pay as he progressed through the ranks. Conversely, it limited pay for black soldiers to ten dollars per month regardless of rank or time in service, meaning the newest recruit received the same amount as a veteran senior sergeant. Thus, African-American soldiers of the Louisiana regiments who fought at Port Hudson and Milliken's Bend as well as those of the Fifty-fourth, who assaulted Fort Wagner, each received three dollars less per month than the army's lowest-ranking white enlistee.

After the formation of the USCT in May 1863, Gen. Lorenzo Thomas requested equal wages for his black soldiers, but the pay scale remained unchanged. The official justification for lower pay for the black regiments came from the preconception that blacks would act only as laborers, building defenses and moving supplies, rather than as combat soldiers. The authors of the act, however, viewed all white soldiers as potential fighters. As a result, whites of the same rank, whether serving at the front in battle or in the rear as administrative support, received the same pay.

Even after the black regiments shed their blood alongside whites in the summer battles of 1863 at Port Hudson and Fort Wagner, payment remained the same. Officials in Washington continued to cite the Militia Act, but blacks and their supporters interpreted the explanation as thinly veiled discrimination against a race considered not as valuable as the white majority.

In protest, some individual black soldiers refused to perform their duties until they received pay equal to that of their white counterparts, and they encouraged their comrades to do likewise. When informed that their protest constituted an act of mutiny, most went back to work. A few still refused, and at least one, Sgt. William Walker of South Carolina, was tried, convicted, and executed by a firing squad.

As a legal means of protest—and more common than refusing orders—black regiments continued to perform their duties but refused to accept any payment at all. Only a few weeks before his death at Fort Wagner, Robert Shaw had suggested to Massachusetts governor Andrew that if his men did not receive full pay, they should be mustered out of service. He wrote, "I shall refuse to have the regiment paid until I hear from you on the subject."

The same abolitionist groups, politicians, and newspapers that lobbied for the black regiments' right to fight joined the protest against unequal pay. Their efforts, and the increased number of Union casualties that necessitated the recruitment of even more African Americans, finally moved Congress to enact legislation on June 15, 1864, providing equal pay. The new act, however, hedged on equality. It provided pay retroactive only to the previous January 1 and limited payment to those black soldiers who were already free on April 19, 1861, when the war began.

Many African-American soldiers did not meet such criteria because they were either runaways or former slaves freed by the advancing Union army. To ensure that all of his men received full back pay, Col. Edward N. Hallowell, Shaw's replacement as commander of the Fifty-fourth Massachusetts, devised a technique of confirming the "free man" status of soldiers in his regiment. In what became known as the Quaker Oath, based on God's law rather than man's, Hallowell allowed each of his soldiers to respond in the affirmative to the statement "You do solemnly swear that you owed no man unrequited labor on or before the 19th day of April, 1861."

Other regiments copied Hallowell's oath to meet the legislative requirements of "free men," and most black soldiers received their pay retroactive to the previous January 1. However, protests about the back-pay issue both in the ranks and by the public continued until Congress, on March 3, 1865, authorized full equal pay retroactive to soldiers' actual dates of enlistment.

Although pay had finally become equitable between black and white soldiers, a vast chasm of discrimination and racism continued to divide the two races both on the battlefield and at home. Fewer than a hundred African Americans, the majority of whom were members of the Louisiana regiments which had evolved from the Native Guards, received commissions as officers during the war. At no time did a black officer ever command white troops.

While denied the rank of officer, black soldiers nonetheless displayed their leadership abilities under fire. At Chafin's Farm, Virginia, on September 29, 1864, four black sergeants took command of their companies after all their white officers had fallen. All four later received the Medal of Honor for their gallant leadership in continuing the attack.

Despite discrimination at all levels, black soldiers, like those at Chafin's Farm, served honorably. With few exceptions, black regiments participated in every battle and campaign during the war's final two years. Desertions were extremely rare, and black soldiers had far fewer incidents of drunkenness and other discipline infractions than their white counterparts. The African-American soldiers willingly assumed the responsibility of proving themselves, freeing their brothers, and preserving the Union. Around their campfires many sang a popular verse:

An African-American guard enforces "riding the sawbuck" punishment at the Provost Marshal's guardhouse in Vicksburg, Mississippi, in 1864. (LIBRARY OF CONGRESS)

So, rally, boys, rally, let us never mind the past;
We had a hard road to travel, but our day is coming fast,
For God is for the right, and we have no need to fear,
The Union must be saved by the colored volunteer.

Despite their stellar performance, the Union army accorded blacks few accommodations. Black soldiers even had to seek their religious comfort and guidance mostly from white officers. Of 139 chaplains assigned to the USCT regiments, only a dozen were African Americans.

In addition to discrimination within the Union army, black soldiers also faced a racist enemy. From the first day they took their place on the battlefield, armed against the Confederate army, black soldiers could expect no quarter for themselves or their officers. In May 1863 the Confederate government formalized these attitudes by approving the death penalty for Union officers leading black soldiers and establishing a policy of enslavement or reenslavement of captured African Americans.

At Fort Pillow, Tennessee, in April, 1864 the most extreme execution of this policy occurred when soldiers under the command of Gen. Nathan Bedford Forrest, a former slave trader and future founder of the Ku Klux Klan, indiscriminately slaughtered at least one hundred blacks and perhaps a few Union officers either during or immediately after their surrender. Accounts by the Confederates claimed that the executions occurred because their commanders lost control of their troops, not because they had given actual orders to kill the prisoners.

The clash became a rallying cry for other black soldiers, who thereafter often went into their own battles shouting: "Remember Fort Pillow." The fight also had its official repercussions. Shortly after the battle, President Lincoln announced that the Union would execute one Confederate officer who was a prisoner of war for each Union officer put to death for leading black troops and that it would sentence one Rebel enlisted prisoner to hard labor for each black soldier subjected to reenslavement. Lincoln's threat, combined with many Southern officers' distaste of the original policy, ended the slaughter and sale of captured members of the USCT. The Rebel army properly treated captured blacks and their officers as prisoners of war for the remainder of the conflict.

At the same time that Lincoln issued his threats, the South began to realize that they were in danger of losing the war and that they, too, might benefit from the military service of African Americans. Some Southern political leaders and military officers, believing that ending slavery to preserve the Confederacy was worthwhile, called for the emancipation of slaves in return for military service. Late in 1863, a group of officers in the Army of Tennessee, led by Gen. Patrick R. Cleburne, drafted a proposal soliciting approval of black recruitment. Their letter, dated January 2, 1864, noted the sacrifices made to date by white Southerners, the contributions of blacks to the Union army, and "the adequate" performance of slaves in previous wars. Within the text, Cleburne stated: "Adequately to meet the causes which are now threatening ruin to our country, we propose . . . that we immediately commence training a large reserve of the most courageous of our slaves, and further that we guarantee freedom within a reasonable time to every slave in the South who shall remain true to the Confederacy in this war. As between the loss of independence and the loss of slavery, we assume that every patriot will freely give up the latter. . . ."

Neither Cleburne's commander nor Confederate president Jefferson Davis accepted the proposal. Gen. Clement H. Stevens best summed up their feelings: "I do not want independence if it is to be won by the help of the Negro. . . . The justification of slavery in the South is the inferiority of the Negro. If we make him a soldier we concede the whole question."

While Southerners were clinging to their concept of black inferiority, African Americans were making significant, albeit indirect, contributions to the Confederacy's war efforts. Slaves were assisting the Confederate army in preparing defenses, taking care of livestock, and serving as general laborers. Black slaves also supported de facto the "cause" by tending farms and working in factories that manufactured war matériel. A few accompanied their owners to the front and directly participated in combat against their would-be liberators. Overall, African-American contributions remained minimal in comparison to their potential—a fact that had seemed tolerable to whites until their "cause" was in dire jeopardy.

While beliefs about African-American inferiority would not change in the white South, the need for their services now convinced more and more Confederates of the necessity of recruiting them. Cleburne died in action at Franklin, Tennessee, on November 30, 1864, but his idea of arming Southern blacks lived on. Confederate secretary of war Judah P. Benjamin wrote on December 21, "The Negroes will certainly be made to fight against us if not armed for our defense. The drain of that source of our strength is steadily fatal, and irreversible by any other expedient than that of arming the slaves as an auxiliary force."

With the Confederacy crumbling on all fronts, Gen. Robert E. Lee made the final plea that moved Davis and the Rebel Congress to authorize black enlistments. On January 11, 1865, Lee wrote to a Confederate senator, "I think, therefore, we must decide whether slavery shall be extinguished by our enemies and the slave used against us, or use them ourselves at the risk of the effects which may be produced upon our social institutions. . . . We should employ them without delay."

On March 13, President Davis signed the Negro Soldier Law, which authorized the enlistment of 300,000 slaves in the Confederate army. The law promised emancipation for honorable service on the condition that slaves' owners and their state governments agree, but a general consensus supported freedom for all who would fight.

The Southern change in attitude was a case of too little, too late. The city of Richmond, Virginia, raised a few companies of African-American Confederate soldiers, but before any large enlistments could occur, Lee surrendered his army, ending the war.

At the same time that blacks in the South played mostly supporting roles as slave laborers at the front, in the factories, and on the plantations, blacks in the North were instrumental in the preservation of the Union and in the emancipation of their race. In doing so, they proved that black Americans could and would fight and earned for themselves a place, albeit far from equal, in the nation's army of the future.

During the Civil War, African Americans represented more than 10 percent of the total Union force. More than 180,000 served in the USCT in 120 infantry regiments, twelve heavy artillery regiments, seven cavalry regiments, and five regiments of engineers. An equal number served in support positions similar to their enslaved Southern brothers. More than twenty-seven hundred African-American soldiers paid the ultimate price by being killed in action. Estimates of black soldiers who died of wounds or from disease and accidents during the long war range from 38,000 to 68,000.

Not only did black soldiers make a significant difference in the strength of the U.S. Army; black sailors played an important role in the operations of the U.S. Navy. When the Civil War began, African Americans, limited by restrictions set by the Navy Department, made up only about 5 percent of the fleet's crewmen. Although their numbers increased significantly, some historians have greatly overestimated the number of black sailors in the Civil War. Most of the confusion comes from a 1902 Naval War Records Office response to a congressional inquiry about black sailors in the Civil War.

The report stated:

> There are no specific figures found in this office relating to the number of colored men enlisted in the U.S. Navy 1861–1865. The total number of enlistments in the Navy from March 4, 1861 to May

Sailors aboard the USS *Miami* repair sails during the Civil War. (ca. 1863). (U.S. NAVAL INSTITUTE)

1, 1865 was 118,044. During the War of 1812 and up to 1860 the proportion of colored men in the ship's crews varied from one-fourth to one-sixth to one-eighth of the total crew. . . . In the absence of specific data it is suggested that as several vessels report during the Civil War having a crew of one-fourth Negroes that the number of enlistments must have been about one-fourth of the total number given above, or about 29,511.

While one-quarter of the crews of several Union ships during the war were black, more recent studies have provided reasonable proof that African Americans made up only about 13 percent of the naval crewmen in 1863, at the height of enlistments of freemen and contrabands. Detailed studies of enlistment records reveal that about ten thousand blacks served at one time or another during the war; that is, about 8 percent of the total force rather than the often quoted 25 percent.

Regardless of the actual numbers, the thousands of African Americans who enlisted in the U.S. Navy during the Civil War served loyally and bravely. While they, like their soldier-brothers, suffered from many discriminatory practices, such as lower pay—receiving only ten dollars per month—black sailors worked in integrated crews and shared equal quarters and rations.

Manpower shortages, always a problem in the navy, provided the impetus to black recruitment. During the first year of the war, the number of U.S. Navy ships increased from 76 vessels to 671, and from 1861 to 1865 the United States commissioned a total of 1,059 ships. Secretary of the Navy Gideon Welles lifted the 5 percent quota on black enlistment on September 20, 1861, but limited their service aboard ship to positions of servants, cooks, and assistant gunners, or "powder boys." Freemen from the North filled most of these expanded billets, but as more ships were launched and more white crewmen succumbed to disease along the Southern blockade, Wells authorized recruitment of contrabands in the spring of 1862. At the same time, he opened all enlisted positions to black sailors but denied them promotion to petty or commissioned officer.

Most ship captains followed Welles's directions and integrated black sailors into all shipboard responsibilities. A few, including Adm. David D. Porter, captor of New Orleans, segregated his crews and barred African Americans from positions such as lookouts because he believed they lacked the intelligence for such responsibilities. Porter employed most of his blacks belowdecks in menial positions. The captain displayed a curious mixture of racism and respect toward his black crewmen, stating, "I cannot get men, so I work darkies. They do first rate work and are far better behaved than their masters."

All across the navy, black sailors proved themselves equal to their tasks. Black sailors Robert Blake, John Lawson, and Aaron Anderson all earned the Medal of Honor for their bravery under fire. A fourth African-American sailor, Joachim Peace, also received the Medal of

Landsman John Lawson: Earned Medal of Honor aboard the USS *Hartford* in Mobile Bay on August 5, 1864. (U.S. NAVAL INSTITUTE)

Honor for his actions as a gun loader aboard the USS *Kearsage* in its sinking of the famous Confederate raider *Alabama* off France's Cherbourg harbor on June 19, 1864.

Another black sailor became one of the war's earliest naval heroes by saving his entire ship. On July 7, 1861, the Confederate privateer *Jeff Davis* captured the cargo-laden Union schooner *S. J. Waring* en route to South America from New York. The Confederates removed the *Waring*'s crew to the *Davis*, except for William Tillman and two others, whom they left as cooks and stewards. They then placed a prize crew of five aboard to sail the captured vessel to Charleston, South

Carolina, where they planned to impress the boat into the Rebel navy and to sell Tillman and his black mates into slavery.

A hundred miles from Charleston, at midnight on July 16, Tillman killed three of the Rebel crew members with a hatchet and captured the other two. With the assistance of his fellow stewards and the captured Rebel sailors, he sailed the ship back to New York, with the U.S. flag waving from the mast.

Less than a year later, still another black sailor delivered a ship to the Union navy, this time one formerly in Confederate service. Early on the morning of May 13, 1862, Robert Smalls and seven fellow slave-sailors of the three-hundred-ton, side-wheel steamer *Planter* waited until white crew members went ashore. They then loaded their own families aboard and sailed out of Charleston harbor. Smalls, a sail maker and ship's pilot, hoisted a white bedsheet and sailed toward the Union blockade offshore, where he delivered the ship and its cargo of supplies and munitions to a Union captain, declaring, "I thought the *Planter* might be of some use to Uncle Abe."

In addition to their freedom, Smalls and his fellow crew members received a prize payment for the ship and its contents. Smalls served for the remainder of the war as a pilot, exploiting his excellent knowledge of the South Carolina coast and intercoastal waterways. After the war he attained an education and in 1876 became a member of the South Carolina Congress, where he served five terms. In frequent speeches, Smalls insisted, "My race needs no special defense, for the past history of them in this country proves them to be the equal of any people anywhere. All they need is an equal chance in the battle of life."

While neither the navy nor the army accepted women of any race into their ranks during the conflict, black women nevertheless contributed to the war effort. In addition to their work and support on the home front, several black women directly served as nurses, scouts, and spies. History credits Mary Louvestre, a slave of an engineer working to convert the captured USS *Merrimac* into the first Confederate ironclad, with stealing a portion of the Rebel plans and delivering them to the Union navy.

Both Sojourner Truth and Harriet Tubman served as nurses in Northern hospitals. Tubman, a former slave and a "conductor" on the Underground Railroad for other runaways, also served briefly as a frontline scout. In June 1863 she accompanied a raid led by Col. James Montgomery up the Combahee River in South Carolina, providing information on the countryside and possible Confederate positions. A later, possibly somewhat inflated account in the Boston *Commonwealth* reported: "Col. Montgomery and his gallant band of 300 black soldiers, under the guidance of a black woman, dashed into the enemy's country, struck a bold and effective blow, destroying millions of dollars worth of commissary stores and cotton . . . and brought off near 800 slaves and thousands of dollars worth of property, without losing a man or receiving a scratch."

Another black woman, Susie King, became a teacher for black soldiers. Because Southern slave laws prohibited the education of blacks, illiteracy hampered them. Once a slave herself, King had learned to read and write in a clandestine Savannah school taught by a free black woman. In 1862, King escaped to the Union-occupied Georgia Sea Islands, where she began her own school and later that year married Edward King. When her husband joined the First South Carolina Volunteers, she went with him and became a laundress, nurse, and literacy instructor for her husband's unit. Susie King remained with the First South Carolina, later redesignated the Thirty-third Regiment USCT, for the remainder of the war.

Following the end of the conflict and the death of her husband, King married Russell Taylor and remained active in Grand Army of the Republic veteran activities. In 1902 Susie King Taylor published her wartime memoirs, which provide the only personal insights into the activities of black nurses and camp followers during the Civil War.

The surrender of Robert E. Lee to Ulysses S. Grant at Appomattox Courthouse, Virginia, on April 9, 1865, and the subsequent collapse of the remainder of the Confederacy over the next few weeks ensured the preservation of the Union and the emancipation of all slaves. African Americans welcomed their freedom and celebrated their participation in actions both on land and at sea that had contributed to the successful outcome of the conflict. However, while they were no longer slaves, blacks were not free from racism, prejudice, and discrimination, which remained prevalent in both the South and the North. Freedom was one thing; equality, another.

In addition to postwar discrimination, African Americans' actual wartime deeds were diminished by white historians, who discounted or ignored the role of blacks. In a 1928 biography of Ulysses S. Grant, white historian W. E. Woodward disavowed any contribution made by black Americans in the Civil War. Woodward wrote: "The American Negroes are the only people in the history of the world, so far as I know, that ever became free without any effort of their own. . . . [The Civil War] was not their business. They had not started the war nor ended it. They twanged banjos around the railroad stations, sang melodious spirituals, and believed that some Yankee would soon come along and give each of them forty acres and a mule."

As well as other white Americans, Woodward dismissed the nearly 200,000 African Americans who served in the war and the more than twenty-seven hundred who died in the fight. Yet blacks had proved to themselves that they could defend their adopted country as well as the white majority—if only provided the opportunity.

Freedom for blacks was the lasting legacy of the Civil War. In terms of military history, the war was also a turning point for African Americans. Black regiments, albeit with white officers, remained on active duty in the postwar army, and black seamen continued to man naval ships. Black and white military history would become inseparable from this point forward.

4

RECONSTRUCTION AND THE INDIAN WARS

Col. Thomas Morgan commanded the Fourteenth U.S. Colored Troops (USCT) from their first action at Dalton, Georgia, on August 15, 1864, until the end of the war. In his memoirs, published in 1885, Morgan wrote about his men's performance at the Battle of Nashville, where more than three thousand Union and six thousand Confederates soldiers were killed, wounded, missing, or captured:

> Colored soldiers had fought side by side with white troops. They had mingled together in the charge. They had supported each other. They had assisted each other from the field when wounded, and they lay side by side in death. The survivors rejoiced together over a hard-fought field, won by common valor. All who witnessed their conduct gave them equal praise. . . . A new chapter in the history of liberty had been written. It had been shown that marching under the flag of freedom, animated by a love of liberty, even the slave becomes a man and a hero.

Because of such valor displayed and casualties taken at Nashville and other Civil War battlefields, the black soldier had earned a permanent place in the military. For the first time in U.S. history, African-American soldiers did not disappear from the ranks with the coming of peace, even when the million-man U.S. Army decreased to an authorized strength of only 54,641 officers and men. In fact, during the postwar reductions, black regiments were some of the last disbanded because they had been the last to join.

The retention of blacks in the army garnered supporters and opponents both within and outside the military. Gen. Benjamin Butler became a vocal advocate, stating that the black soldier had "with bayonet . . . unlocked the iron barred gates of prejudice, and opened new fields of freedom, liberty, and the equality of right."

Others, such as Sen. Willard Saulsbury of Delaware, did not share Butler's sentiments. To his fellow U.S. senators, Saulsbury questioned, "What would be the effect if you were to send Negro regiments into the community in which I live to brandish their swords and exhibit their pistols and their guns?"

After much debate, an assignment process evolved that generally met the desires of both sides. Black soldiers remained in the army but received postings to remote areas west of the Mississippi where they would have limited contact with white Americans.

The only exception to the frontier assignments for black soldiers was duty as part of the occupation forces in the former Confederate states. Some Northern politicians and military officers relished the punitive impact of stationing blacks in the Deep South, where the two colors that could inflict the most pain and anguish were the blue uniform tunics of the Union army and the black skin of former slaves now soldiers wearing those uniforms. As a result, African Americans served as part of the occupation forces in several Southern states immediately after the war.

Reconstruction administrators from the North, known as carpetbaggers for their hastily packed luggage, made of carpet material, controlled most southern state and municipal governments and militia positions during the period. While evidence indicates that many of these officials were far more concerned with acquiring personal wealth and status than with ensuring proper conduct of their office, their tenure did allow limited opportunities for some blacks to flourish in both politics and in the state militias. Robert Smalls, the former captain of the *Planter*, gained promotion to major general in the South Carolina militia, and several other former slaves and freemen secured militia commissions and political office.

White-robed members of the Ku Klux Klan and similar organizations waged a reign of "bullet and the noose" terror to put their former chattel into "their place" by excluding or limiting their access to voting, education, and every other aspect of equality. Throughout the South, and indeed the North as well, the United States reverted back to a segregated society in which whites denied blacks equal rights in the country they had fought to preserve. Even Smalls, who remained in politics, became much better known to white South Carolinians as "the boat thief" for his delivery of the *Planter* to the Union navy rather than a war hero or militia officer.

Some, however, did not waver. The "liberal Republicans" in Congress demanded that blacks be a part of the postwar Regular Army when Congress began debate about the future of the military in March 1866. A compromise on retaining African Americans in the army was ultimately won by the Republicans, who believed that blacks had earned the right to serve, and by pragmatic representatives, who knew that the hardships and dangers of the Western plains would limit the number of volunteers.

On July 28, 1866, a congressional act authorized sixty-seven regiments—five artillery, twelve cavalry, and fifty infantry—to make up the Regular U.S. Army. Early drafts of the act called for six of the infantry regiments to be manned by blacks, but lobbying by Benjamin F. Wade of Ohio produced authorization for two of the six black regiments to be cavalry. Although African-American soldiers had successfully manned artillery units in the Civil War, advocates lost their bid to open that branch of the service to blacks because too many congressmen believed that African Americans did not possess the intelligence to be artillerymen. They led the vote to restrict the black regiments to the cavalry and infantry.

During the organizational and training phase for the Ninth and Tenth Cavalry Regiments and the Thirty-eighth, Thirty-ninth, Fortieth, and Forty-first Infantry Regiments, Congress further reduced the number of regiments, black and white, throughout the army. While the Ninth and Tenth Cavalry remained intact, the four infantry regiments merged into two. The Thirty-eighth and Forty-first consolidated into the Twenty-fourth Infantry Regiment and the Thirty-ninth and Fortieth became the Twenty-fifth Infantry Regiment. The U.S. revised statutes of 1869 designated all four regiments as "colored."

Despite their distinguished service in the Civil War and the professionalism of enlisted men reporting to the new black regiments, African Americans in uniform suffered the same discrimination and racism as their civilian brothers. Along with prejudice against the color of their skin, African-American soldiers also experienced the disdain that their white countrymen felt toward soldiers in general. While lauded as saviors during the Civil War, all soldiers in the postconflict years—regardless of color—were viewed as a burden on the economy and as men who could not find other employment.

The black man in uniform found himself at the bottom of the despised military group. White Americans in both the North and the South generally condemned the black soldiers as shiftless, lazy, undisciplined, and ignorant. As they had proved in the Civil War and would continue to display in the long decades of the Indian Wars, African-American soldiers were guilty of none of these stereotypical accusations. The only description remotely accurate was "ignorant," in the literal sense, because most black enlistees of the period could neither read nor write. This, however, was not a matter of intelligence but the result of prewar restrictions in the South against educating African Americans.

Despite the lack of social esteem afforded soldiers in the postwar years, African Americans found many advantages in the military. The opportunity for education was one. Exposed to the classroom, usually headed by the regimental chaplain, blacks displayed a yearning for education and an ability to learn equal to their white counterparts. Another appeal was that since the prosperity of the postwar years did not otherwise extend to blacks, the military provided one of the few semblances of equality. In addition to food, clothing, and shelter, each recruit—regardless of race—received the same thirteen dollars per month salary, with an annual increase of one dollar per month and a reenlistment bonus at the end of five years.

While white regiments had difficulty in finding and retaining volunteers, black regiments had more applicants than positions available. Whereas white recruits believed that joining the army represented a step down in social position, black volunteers viewed military service as an elevation in status and an opportunity to improve themselves.

Both races shared rather simple enlistment requirements. Recruiters required volunteers to be physically healthy, unmarried, and between the ages of eighteen and thirty-five. Literacy was not a prerequisite. The army kept no statistics at the time, but the study of regimental records reveals that the typical black volunteer was about twenty-three years of age, either a farmer or laborer, and illiterate.

Filling the ranks of the four postwar black regiments proved to be no difficulty, but finding white officers to lead them created problems. Military personnel and civilians viewed officers assigned to black regiments as inferior to those serving with white units. As a result, officers often turned down commands as captains in the black regiments and accepted positions as lieutenants in white regiments. Few graduates of the U.S. Military Academy sought service in black regiments, making the artillery, which did not accept black enlistees, the most popular branch among West Point graduates. A young officer at Fort Monroe, Virginia, in 1883 perhaps best summed up the feelings of many of his contemporaries when he said that he would "rather be a second lieutenant of artillery than a captain of niggers."

While few desired to serve in black regiments, most white officers firmly believed that African Americans were incapable of leading themselves. Authorization of the four black regiments neither required whites in command positions nor excluded blacks officers from serving in any position. There were, however, no black officers available to seek positions in the regiments. The few African Americans who had gained commissions during the Civil War had been discharged after Appomattox. In the postwar era, blacks could apply for commissions, but in every instance their applications were denied.

With no commissions available from the ranks or by direct appointment from civilian life, the only other avenue for a black man to become an officer was the U.S. Military Academy. However, West Point, since its establishment in 1802, had never accepted an African American into its corps of cadets. In 1870, South Carolina congressman Solomon L. Hoge appointed James Webster Smith, a student at Howard University, as the academy's first black cadet. Smith passed the mental and physical entrance requirements but never gained the acceptance of his white classmates.

For the next four years, Smith experienced both official and unofficial harassment. Twice court-martialed, once for an altercation with a white cadet and a second time for "conduct unbecoming a gentleman," Smith endured being sentenced to repeating his plebe year. During his entire stay at West Point, Smith experienced ostracism by the white cadets; they "silenced" him, refusing to speak to him except in the line of duty. Smith remained resilient for four years before the academic department declared him deficient in a philosophy course and dismissed him from the academy in June 1874.

In 1873, a year prior to Smith's dismissal, a second black cadet, Henry O. Flipper of Georgia, entered the academy. Although he suffered the same "silence" treatment for four years, Flipper endured and in 1877 became the 2,690th graduate of the U.S. Military Academy and its first African-American alumnus.

Flipper's treatment improved little after he joined the Tenth Cavalry, where he became the sole black officer not only in the regiment but in the entire U.S. Army. From all evidence, Flipper performed well for four years before being court-martialed for the alleged embezzlement of two thousand dollars from commissary funds. Although acquitted of the primary charge, the board convicted him of "conduct unbecoming an officer" and discharged him. Flipper maintained at his trial and for the rest of his life that his fellow officers had conspired against him first because of his race and, second, because of attentions he had paid to a white woman at Fort Concho, Texas.

Despite his later success as a civilian engineer, Flipper carried the stigma of the army's charges. Not until nearly a century later did the U.S. Army Board for the Correction of Military Records review the circumstances of Flipper's discharge. The result was to clear his name and award him an honorable discharge. On the hundredth anniversary of his graduation from West Point in 1977, the academy dedicated a memorial bust in the cadet library to Lieutenant Flipper.

From the end of the Civil War until the turn of the century, only twenty-five black men received appointments to West Point. Of this number twelve passed the entrance examination. In addition to Flipper, only two others, John Alexander of Ohio, in the class of 1887, and Charles Young of Ohio, in the class of 1889, graduated and received commissions as second lieutenants in the Regular Army.

Discrimination against blacks in the military was by no means limited to black West Point cadets and graduates. The black regiments received no better welcome from the army. The Fortieth Infantry Regiment, soon to be redesignated the Twenty-fifth, assembled at the Goldsboro, North Carolina, railway station on March 31, 1869, only to find freight and cattle cars rather than passenger trains for their ten-day trip to Louisiana. When the Tenth Cavalry Regiment assembled at Fort Leavenworth, Kansas, in 1867, the post commander assigned them a swampy area for their camp, did not allow them to march in reviews, and ordered that no black cavalryman come within ten yards of a white soldier.

The Tenth Cavalry, as well as the other three black regiments, also suffered discrimination in the issuance of arms, horses, and supplies. In evaluating the first fifty horses assigned to the Tenth Cavalry, white commander Col. Benjamin Grierson reported that not a single animal was suitable for service. Included in the herd were cripples and worn-out horses more than a dozen years old that were veterans of the Civil War. The black regiments also received the lowest priority in the issue of saddles, harnesses, weapons, and uniforms.

The horse and equipment situation did not improve with time. In a letter to Colonel Grierson on May 22, 1870, the commander of Company H, Tenth Cavalry, Capt. Louis H. Carpenter complained that his unit was "getting mean and worn out horses from the 7th Cavalry. Since our first mount in 1867 this regiment has received nothing but broken down horses and repaired equipment."

Perhaps the most pronounced indicator of the War Department's neglect of the black regiments was the lack of symbols for the units themselves. Historically, every regiment in the U.S. Army, as well as most military units around the world, has always carried a unit flag, usually an ornately embroidered silk, known as "the colors." These standards symbolize the identity of a regiment. They have been the impetus for some of the most remarkable examples of bravery in battle when soldiers fought to defend their colors from capture.

Yet the army did not issue the professionally manufactured regimental flags to the black units. So black soldiers secured cloth, thread, and needles to sew their own. Although not as "pretty" as the flags of the white regiments, the black soldiers served proudly under colors they created in their own barracks until the official flags finally arrived.

In addition to suffering the lowest priority for everything within the army, black regiments had to contend with hostile attitudes toward them in the towns and territories to which they were assigned. Even in the far reaches of the frontier, white pioneers refused to accept black soldiers as equals. The only places they found a welcome were "hog ranches," the separate areas just outside of town where saloons and brothels provided entertainment to the black soldiers on payday as long as their meager wages lasted.

Although they were protecting the settlements, black soldiers could not depend on equal justice from white sheriffs in the towns adjacent to their forts. On August 25, 1885, whites broke into the Sturgis City Jail in Dakota Territory and lynched a soldier of the Twenty-fifth Regiment charged with murdering a white man. At Sun River, Montana Territory, on June 10, 1888, vigilantes removed another member of the regiment from the local jail and lynched him. None of those responsible for either lynching was ever brought to justice.

Even when apprehended, whites seldom faced punishment for murdering black soldiers. On January 31, 1881, white sheep rancher Tom McCarthy shot and killed am unarmed private, William Watkins of Company E, Tenth Cavalry, in a San Angelo, Texas, saloon. Members of the Tenth Cavalry at nearby Fort Concho captured McCarthy and turned him over to the San Angelo sheriff, who released the rancher pending an investigation.

The African-American soldiers of the Tenth Cavalry did not take kindly to the murder of one of their comrades and the release of his killer. A handbill dated February 3, 1881, and signed by "U.S. Soldiers" appeared throughout the town: "We, the soldiers of the U.S. Army, do hereby warn the first and last time all citizens and cowboys, etc., of San Angelo and vicinity to recognize our right of way as just and peaceful men. If we do not receive justice and fair play, which we must have, some one will suffer—if not the guilty, the innocent. It has gone too far, Justice or death."

Some of the black soldiers followed the announcement by riding into San Angelo and firing warning shots into several buildings before Grierson confined the regiment to the post. Ultimately, a grand jury indicted McCarthy for first-degree murder, but a jury in Austin, Texas, took only minutes to find him not guilty.

Beyond the problems with supplies and the poor treatment by the civilian population they protected, black soldiers also faced unrelenting natural hardship in the vast western frontier. The white regiments, which shared frontier duties, rotated back East to relatively easy assignments in fairly pleasant surroundings every several years. The four black regiments, however, in what apparently was an "out of sight, out of mind" policy, remained on continuous duty in the West for more than two decades.

The black infantry and cavalry regiments served at remote outposts in extremely harsh conditions. An 1872 inspector general's report described Fort Clark, Texas: "The quarters are wretched and therefore nothing beyond shelter and ordinary police can reasonably be looked for. All except two companies of cavalry are in huts. . . . There is no place for divine service or instruction. . . . The public stores are imperfectly covered with paulins [*sic*] or put in insecure huts, improvised at great expense."

When the Twenty-fifth Infantry Regiment moved from Texas to Fort Randall, Dakota Territory, they found conditions improved, but a report in 1881 noted that the barracks were overcrowded, short of furniture and lighting, and generally in need of repairs. At nearby Fort Hale, black soldiers fared no better. An inspection report of the black regiment made by Capt. R. P. Hughes of the white Third Infantry Regiment in October 1883 noted: "The post commander has his post in excellent condition as to cleanliness and order, but he cannot make a tumble down log hut look either beautiful or inviting. I respectfully submit to the Department Commander that these two companies are so much more badly off than any of the other troops in this geographical department that I think some strong measures should be adopted for their relief. . . . The troops are good, but their accommodations are wretched."

As if it were not enough that they suffered a low priority in receiving matériel, general discrimination by the white settlers, and wretched living conditions on their posts, African-American soldiers also faced one more major complication. The black regiments reported to the western frontier at the height of the Indian Wars, which reached their bloodiest period in the post–Civil War years.

It was, however, from the Indian enemy that the black regiments gained their most enduring label. While whites, whom they protected, referred to the black soldiers as "coloreds," "brunettes," "Africans," and much more commonly as "niggers." Indians, who faced them on the battlefield, had a more complimentary name for them—Buffalo Soldiers. Whether the name began as a form of respect or simply an acknowledgment of the similarity between the curly black hair of African Americans and that of the buffalo is unknown. Whatever the source, black soldiers, appreciating the comparison to the stoic animals the Indians considered sacred, adopted the name. The Tenth Cavalry even added a buffalo to their unit crest.

During the Civil War, Native American tribes on the Plains and in the Southwest had reclaimed much of their lands after army units withdrew east to fight the rebellious Confederates. When the war ended, veterans of both sides headed west in search of new opportunities, only to find the Indians prepared to fight to retain their way of life. For the next quarter century, much of the protection for white expansion westward came from the black infantry and cavalry.

For their first eight years, the Tenth Cavalry served in the Central Plains and then transferred to Texas in 1875, replacing the Ninth Cavalry, which moved to the New Mexico and Arizona Territories. The Twenty-fourth and Twenty-fifth Infantry Regiments also served in Texas. In 1880 the Twenty-fourth moved to the Oklahoma Territory and then, eight years later, on to the New Mexico and Arizona Territories.

During their service in Texas, the Twenty-fifth recruited yet another group of blacks to assist them—the only other black unit to participate in the Indian Wars besides the original four regiments. These new recruits were descendants of African Americans and Seminole Indians who had migrated to Mexico at the end of the Second Seminole War in Florida and settled near the Texas border.

Because neither soldiers nor officers of the Twenty-fifth had experience in the southwestern terrain when the unit arrived in Texas, the regiment recognized its need for assistance from someone familiar with the territory. In 1870, Maj. Zenas R. Bliss crossed the border and convinced about fifty of the racially mixed black Seminoles to serve as scouts in return for land and food in Texas. The Seminole Negro Indian Scouts, under command of white lieutenant John Bullis, served for nine years with the Twenty-fifth Regiment and occasionally assisted other black regiments. The scouts participated in twenty-six major expeditions, including several raids into Mexico in pursuit of renegade Indians and bandit gangs.

Although these raids proved successful in halting the incursions from Mexico into Texas, they also engendered political difficulties and racial prejudice. After the Seminole Negro Indian Scouts assisted the Twenty-fifth Regiment in the summer of 1876, Gen. E. O. C. Ord, the commander of the Department of Texas, wrote to his superiors: "I must remark, however, that the use of colored soldiers to cross the river after raiding Indians, is in my opinion, impolitic, not because they have shown any want of bravery, but because their employment is much more offensive to Mexican inhabitants than white soldiers."

Despite the criticism, Seminole Negro Indian Scouts provided a needed service and, from all accounts, they performed well; four of them received the Medal of Honor. While the scouts more than met their end of the agreement by assisting black regiments in Texas, neither the army nor the U.S. government kept the promise of food and land grants. A *New York Times* reporter noted on September 1, 1875, that the scouts "are now living in great destitution, bordering on starvation."

Bliss, Bullis, and other officers lobbied for better treatment of the Scouts to no avail. When the border area became peaceful in 1881, the Seminole Negro Indian Scouts were disbanded and melted into oblivion.

The Twenty-fifth Regiment left Texas in 1880 without its Seminole Negro Indian Scouts for the Dakota and Montana Territories, but only after much debate on whether the black soldier could perform in a cold climate. Many officers, especially those in Washington, D.C., who had never served in African-American regiments, thought that because blacks had originated in the African tropics and then lived mostly in the American South, they would not be able to endure the harsh winters of Montana and the Dakotas. The army's quartermaster general expressed concern, stating, "Colored men will not enlist with the prospects of going to that rigorous climate . . . the effect of the cold will be very injurious to those men whose terms of enlistment do not soon expire."

As usual the Buffalo Soldiers proved their critics and doubters wrong. The infantrymen of the Twenty-fifth Regiment found the northern Rockies a welcome relief from Texas and enjoyed an improvement in barracks and facilities. As for the cold, blacks showed that they could soldier under any condition. In the spring of 1881, after an extremely severe winter, black soldiers performed relief operations, rescuing stranded white homesteaders and their animals and delivering supplies to remote villages and camps. The regiment's Company F alone assisted more than eight hundred settlers in Dakota Territory, and when government supplies ran low, they contributed their own pay to provide additional relief supplies.

Despite their many missions, the primary purpose in assigning black regiments to the West was to suppress hostile Indians. African-American soldiers participated in some of the earliest fights against the Native Americans in the post–Civil War period and continued their campaigns until the concluding battles of the Indian Wars. The first skirmish occurred only a few months after the establishment of the Tenth Cavalry Regiment. On August 2, 1867, thirty-four members of the regiment's F Company rode in pursuit of a Cheyenne war party that had raided a camp of railway workers near Fort Hays, Kansas.

Capt. George Armes, in command of the F Company troopers, followed the Indians' trail, only to be attacked and surrounded by a group of seventy-five to eighty warriors. After a six-hour fight in which they killed six Indians, the cavalrymen broke through the encirclement. During the withdrawal, the Tenth Cavalry suffered its first fatality of the Indian Wars when Sgt. William Christy of Pennsylvania fell from a gunshot to the head.

A year later, the Tenth Cavalry rescued a detachment of scouts under the command of Maj. George A. Forsyth from a siege by Cheyenne, Sioux, and Arapahos on an island in the Republican River in eastern Colorado. Moving day and night by alternating riding and walking to rest their horses, the soldiers of H Company rescued the white scouts after an eight-day journey.

Over the next twenty-five years, the Ninth and Tenth Cavalry and the Twenty-fourth and Twenty-fifth Infantry Regiments served throughout the West and Southwest, their assignments based on wherever the most significant Native American threat loomed at the time. During that period, the regiments engaged in more than two hundred battles and skirmishes, and their soldiers received eighteen Medals of Honor for bravery under fire.

The first Medal of Honor earned by an African American in the Indian Wars typified both the personal bravery and style of operations of the black regiments. On the morning of May 20, 1870, a ten-man patrol under the leadership of Sgt. Emanuel Stance from Company F, Ninth Cavalry Regiment, departed Fort McKavett in southwest Texas to scout twenty miles north to Kickapoo Springs. About halfway there, they spotted a half-dozen Indians driving a herd of horses suspected to be stolen. Stance, a native of Charleston, South Carolina, and barely five feet tall, ordered a charge that resulted in a running fight for eight miles that dispersed the Indians and captured nine horses.

Ninth Cavalry Regiment at Fort DuChesne, Utah, in 1897. (U.S. ARMY MILITARY HISTORY INSTITUTE)

Stance and his patrol rode on to Kickapoo Springs, where they camped for the night. On their way back to Fort McKavett, the cavalrymen surprised twenty Indians stalking a herd of horses and its small guard detail. Stance and his men again charged, dispersed the Indians, and captured five more horses. Minutes later, the Indians counterattacked, but Stance personally flanked the Indians and forced their retreat. Stance modestly reported on May 26 that when the fight began, he had "turned my little command loose on them . . . and after a few volleys they left me to continue my march in peace." His superiors recognized his bravery with the Medal of Honor.

The Native Americans rarely assembled numbers large enough to challenge Regular Army units and only on a few occasions showed any inclination to do so. Most of the Indian attacks focused on isolated ranches or on small parties moving cross-country. Guerrilla warfare—hit-and-run raids and small skirmishes of short duration—characterized the Indian Wars.

As a result, U.S. Army regiments on the frontier spent much of their time attempting to find Indian camps or pursuing raiders who had attacked farmers, miners, or other workers. Because of its mobility, the cavalry spent more time than the infantry in the field, covering vast distances looking for the hostiles.

After a month-long campaign by two companies of the Ninth Cavalry across West Texas in April–May 1870, the operation's commander, Maj. Albert P. Morrow, wrote in his official report, dated June 1, 1870, that his black cavalrymen had

> marched about 1,000 miles, over two hundred of which was through country never explored by troops, drove the Indians from every rancheria . . . destroyed immense amounts of . . . food, robes, skins, utensils and material, and captured 40 horses and mules. I cannot speak too highly of the conduct of the officers and men under my command, always cheerful and ready, braving the severest hardships with short rations and no water without a murmur. The Negro troops are peculiarly adapted to hunting Indians, knowing no fear and capable of great endurance.

The infantry regiments drew guard duty at the forts and at key points along the various communications routes, where they also on occasion became engaged with hostile Indians, who attacked work parties, mail stages, or supply trains. Besides pursuing their adversaries and guarding areas such as railway and telegraph construction sites, black regiments also performed routine garrison duties, which often included the building of their own forts and quarters and the care of their equipment and animals.

Despite all the hard work, the harsh weather, and the dangers from hostiles, boredom proved to be the enemy most difficult to combat. The remoteness of the camps and forts, combined with a lack of recreational facilities, frequently led to trouble. The African-American cavalrymen and infantrymen in the West proved that soldiers will be soldiers regardless of color. They will perform brave deeds when called upon; they will also get into trouble when left idle.

The ranks of the Buffalo Soldiers included deserters, malingerers, drunks, thieves, and even the occasional murderer. They were far from perfect, yet in every statistical category their meritorious performance exceeded that of the white regiments. In 1889, Secretary of War Redfield Proctor noted in his annual report to Congress that the desertion rate in the black regiments was only 2 percent, compared to 12 percent in the white units. Black regiments also had fewer alcohol-related incidents and enjoyed higher reenlistment rates.

Many direct and indirect factors influenced the superior morale and performance of the black regiments. Positive attributes—unit pride, a sense of belonging, and feelings of accomplishment—all contributed to the low desertion rate, as did the knowledge that each black soldier represented his entire race and determined its future treatment both in and outside the military. At the same time, negative alternatives also played a role in preventing black soldiers from deserting. Despite the hardships of their life, the army offered far more opportunity than did the civilian world for African Americans. Moreover, a black soldier on the run stood out anywhere he went in the predominantly white western territories.

By the end of the 1880s white farmers and ranchers occupied much of the West, and the various renegade Indians were either dead or restricted to reservations. However, black regiments continued to perform escort duty along the various stage and supply routes and to combat the final Indian uprisings. Their performance remained as dedicated in the final days as it had been during their more than two decades of service on the frontier.

On March 7, 1890, only ten days before the formal end of the campaigns against the Apaches, Sgt. William McBryar of Elizabethtown, North Carolina, earned the Medal of Honor near Fort Thomas, Arizona. The Tenth Cavalry soldier exposed himself to return fire against an Apache ambush that saved his comrades. In an understated nomination for the award, 1st Lt. J. W. Watson wrote, "Sergeant McBryar demonstrated coolness, bravery, and good marksmanship under circumstances very different from those on the target range."

On May 11, 1889, an eleven-man escort, composed of soldiers from the Twenty-fourth Infantry, engaged in their last major fight in the Southwest, on the road between Cedar Springs and Fort Thomas, Arizona, against bandits attempting to rob an army paymaster wagon. Two of the black soldiers, Sgt. Benjamin Brown of Spotsylvania, Virginia, and Cpl. Isaiah Mays of Carters Bridge, Virginia, received the Medal of Honor, and seven others were awarded Certificates of Merit for their valor.

In his official account, Maj. J. W. Wham, paymaster and commander of the detail, reported, "I served in the infantry during the entire Civil War and served in 16 major battles, but I never witnessed better courage, or better fighting than shown by those colored soldiers."

The final Medal of Honor awarded to an African American during the Indian Wars came in the conflict's last campaign. During the Sioux "Ghost Dance" uprising in late 1890, the Ninth Cavalry supported the white Seventh Cavalry in pursuing and returning Big Foot and his followers to their reservation. On December 13, near Pine Ridge, South Dakota, a patrol of I Company, Ninth Cavalry, found themselves surrounded by a large Sioux force. Cpl. William O. Wilson fought his way through the encirclement to bring help which rescued his fellow troopers.

During their more than two decades of service, generally in the most adverse terrain and against the most hostile Indians, the four regiments of black soldiers earned the respect of their Native American adversaries, the white regiments with whom they campaigned, and finally, Washington officials.

In recognition of their long and faithful service, the War Department retained all four black regiments on active duty after the conclusion of the Indian Wars. The battalions of the Ninth and Tenth Cavalry and the Twenty-fourth and Twenty-fifth Infantry were spread across the West, performing garrison duties and maintaining combat readiness. As an additional reward for their Indian War service, Secretary of War Redfield Proctor ordered the transfer of K Troop, Ninth Cavalry, from Fort Robinson, Nebraska, to Fort Myer, on the edge of Washington, D.C. They arrived on May 25, 1891, and for the first time since Reconstruction, black soldiers were serving east of the Mississippi River.

The Ninth Cavalry troopers remained for three years in Washington, where they gave riding demonstrations and performed ceremonial duties, including acting as a special escort for President Cleveland during a parade in 1893. Although their performance both on and off duty was exemplary, some white residents of Washington as well as Southern members of Congress complained about the proximity of the black cavalrymen. On October 3, 1894, K Troop returned to Fort Robinson and rejoined their regiment.

During their tour in Washington, all of K Troops' facilities—billets, dining halls, the post exchange, and the barbershop—were segregated. General discrimination remained a fact of life for the black regiments no matter where they served. Shortly before K Troop rejoined the Ninth Cavalry at Fort Robinson, a reporter from a black Ohio newspaper visited the Nebraska camp. His report in the July 14, 1894, edition of the *Cleveland Gazette* declared:

> There is considerable discrimination going on at Fort Robinson, Neb. There are three white clerks in the commissary department, two in the post exchange, two in the post bakery, two in the post adjutant's office, two in the officers' club and mess room, the post librarian is a white soldier, two white soldiers at the post pump house, a white non-commissioned officer in charge of the post saw mills, five white men in the post quartermaster's department. All these places are filled by enlisted men of the Eighth Infantry of which there are only two companies at the post, while there are six cavalry companies, all colored.

Although all African Americans had gained their freedom in 1865 and the black regiments had performed well for more than two decades during the Indian Wars, the overall treatment of blacks had deteriorated by the close of the nineteenth century. Racial segregation, as established by what became known as Jim Crow laws, named after a popular contemporary minstrel song, separated black and white Americans in every way. Poll taxes, literacy tests, and other similar measures kept the right to vote from most blacks in the Southern states.

Various local and state segregation laws received federal support from the May 18, 1896, Supreme Court decision in the case of *Plessy v. Ferguson*. Although the issue focused on segregated public transportation in Louisiana, the court's majority opinion stated that both races benefited from segregation: "Legislation is powerless to eradicate racial instincts or to abolish distinctions based upon physical differences, and the attempt to do so can only result in accentuating the difficulties of the present situation."

At the end of the nineteenth century, Jim Crow laws in the South barred or restricted African Americans from white schools, churches, hotels, restaurants, theaters, and most every other public facility. Whites often enforced these laws by the hangman's rope or by fire. In both 1892 and 1893 white Americans either lynched or burned alive more than 150 black Americans.

Within the army, the four black regiments continued to be commanded almost exclusively by white officers and to face the same segregation as did their civilian brothers. Despite these conditions, the U.S.

Army did provide equal pay and a semblance of equal, albeit separate, facilities. At the close of the nineteenth century, the army continued to be one of the few U.S. institutions to afford African Americans the opportunity of service, advancement, and accomplishment.

Likewise, black sailors were afforded these opportunities but had to endure the same discrimination as those of the black army regiments and the black civilian population. Despite the reduction in ships and sailors in the post–Civil War period, the navy continued to enlist African Americans because it was still difficult to fill crews. During the first thirty years after the Civil War, the U.S. Navy averaged five thousand to six thousand men on duty at any given time, of whom 10–14 percent, or five hundred to eight hundred, were black. Because of the limited space aboard ships, both sleeping and eating areas were integrated.

Naval regulations did not restrict blacks from becoming regular seamen or gunners. In 1870 only 29 percent of black sailors served in the "domestic" positions of cooks and stewards, but this number increased to 49 percent by 1890. No official policy change prompted this movement, but some naval leaders believed that African Americans did not have the intelligence to adapt to the intricacies of the steam boilers and mechanization that replaced most of the old sailing ships in the 1880s. Since they had the latitude to transfer crew members wherever they wished, many captains chose to place the blacks in kitchens and dining rooms.

Despite mounting discrimination, black enlisted men served the navy bravely. Seaman Joseph B. Noil, who was born in Nova Scotia and enlisted in New York City, was the first African American to receive a post–Civil War Medal of Honor, by saving the life of a fellow sailor who fell overboard from the USS *Powhatan* on December 26, 1872. Five other black sailors received the medal before the end of the century for similar rescue efforts. One, Seaman Robert Sweeney, received the award twice for saving drowning sailors while aboard the USS *Kearsarge*, on October 26, 1881, and the USS *Jamestown*, on December 20, 1883.

Racism rather than recognition, however, remained the norm. When abolitionist leader Frederick Douglass was appointed U.S. minister to Haiti in 1889, the navy was ordered to transport him to his new post. The first captain given the assignment resigned, the second claimed

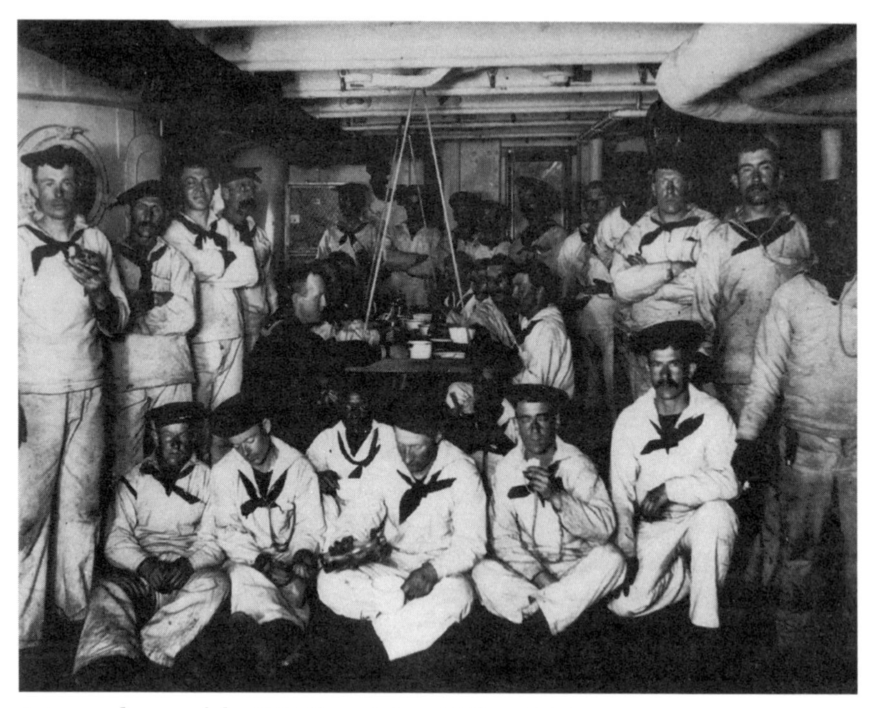

Integrated crew of the USS *Newark* (ca. 1895). (U.S. NAVAL INSTITUTE)

his ship unseaworthy for the voyage, and the third requested a transfer to a new command.

Discrimination against African Americans was also blatant in the selection of officers. Between the Civil War and the end of the century, no African American gained a commission in the U.S. Navy. The navy continued to limit the number of black naval volunteers and barred them from direct commissions. The only other avenue for commissions was the U.S. Naval Academy at Annapolis, Maryland, which proved to be just as inequitable as West Point—in fact, more so.

John Conyers of South Carolina received the first nomination of an African American to the U.S. Naval Academy in 1872, three decades after the school's establishment in 1842. Conyers found himself welcomed by neither faculty nor fellow cadets and suffered the same "silence" treatment that West Point cadets inflicted on black students. After a year of enduring ostracism and hazing, Conyers resigned when declared deficient in his mathematics and French classes.

Messman in the wardroom pantry of the USS *Brooklyn* (ca. 1896).
(U.S. NAVAL INSTITUTE)

A second black cadet, Alonzo McClennan, reported to Annapolis in September 1873, followed a year later by Henry E. Baker Jr. Both encountered the same treatment that had greeted Conyers. McClennan resigned after six months; Baker stayed only ten weeks. No other African American received an appointment to the U.S. Naval Academy until 1936—more than seventy years after the end of the Civil War and sixty years after the resignations of the first three black cadets. It would be another thirteen years before the first black graduated and received his commission from the academy in 1949.

The more than a quarter century following the Civil War proved that segregated black cavalry and infantry units could perform as well as, or better than, white regiments. Sailors, integrated into ships throughout the navy, displayed their ability to contribute as efficient crew members. Soldiers and sailors alike set the example as patriotic black Americans who could and would serve loyally and bravely, if only given the chance. They were soon to receive even more opportunity to fight for their country and to continue to display their value to the American military.

5

THE SPANISH-AMERICAN WAR: CUBA AND THE PHILIPPINES

During the brief period of peace after the official end of the Indian Wars in 1891, African Americans in the U.S. military suffered increased discrimination and racism. With black soldiers no longer needed to man remote, dangerous outposts in the West and as more whites volunteered for naval service, many Americans resumed their unfortunate but predictable peacetime opposition to blacks in the military. However, before either army or navy officials could decide the future of African Americans in the services, the battleship USS *Maine* exploded and sank in Havana harbor on February 15, 1898, setting off yet another war.

For a decade the United States had been demanding that Spain improve its treatment of its colonials in Cuba, and many Americans were in support of local revolutionary efforts to achieve independence. Sailing the uninvited *Maine* into Havana had been a U.S. ploy to show approval of a tentative compromise between the Spanish government and the Cuban rebels. Instead of solidifying peace, however, it started a war between the United States and Spain.

Future investigations would determine that the *Maine* exploded as the result of internal combustion in its coal bunkers. However, it was initially assumed that a Spanish mine or torpedo had struck the ship. In an era of competitive yellow journalism, U.S. newspapers inflamed the public with such jingoistic slogans as Remember the *Maine*. When diplomatic attempts to resolve the crisis peacefully failed, the United States declared war against Spain on April 21, 1898.

Rabid patriotism swept the country, prompting citizens from every class and social station to volunteer for military service. Thousands of white and black men came forward to enlist in the U.S. Volunteer Infantry (USVI) and state militias formed in response to President William McKinley's call for 125,000 volunteers on April 23 and for 75,000 additional recruits on May 26. Although they signed up in sufficient numbers, the volunteers had to be organized and trained before transfer to the battle zone.

In the meantime, the war effort fell on the shoulders of the Regular Army units. Acting on the commonly held belief that blacks were inherently better suited for the tropical climate and more resistant to its diseases than whites, the War Department had begun transferring the Regular Army black regiments to staging areas in the South even before the formal declaration of war.

In March 1898 the army ordered the all-black Twenty-fifth Infantry Regiment from Montana to Dry Tortugas Island, between Key West, Florida, and Cuba. Orders for the other black regiments followed on April 14, directing the Ninth Cavalry from Nebraska and Tenth Cavalry from Montana to Chickamauga Park, Georgia. Simultaneously, the Twenty-fourth Infantry in Utah received orders to transfer to New Orleans. The regiments departed from their home stations to the music of bands and general support from white residents.

In Saint Paul, Minnesota, the local population assembled at the train station to cheer the flag-draped train carrying the black soldiers. The farther south the black regiments traveled, however, the more the welcome cooled. Black men in blue uniforms remained unacceptable to most Southerners, who showed their racist displeasure by enforcing the same Jim Crow segregation laws on soldiers as they did on civilian African Americans. Black soldiers, well aware that their four regiments were among the finest in the army, resented the discrimination. A correspondent aptly summarized their feelings in a May 1898 dispatch to the *Boston Evening Transcript*: "The Negro soldiers who come here from the North and West . . . are under the impression that they are as good as white soldiers . . . they think that the willingness to die on an equality with white men gives them a claim to live on something like an equality with them."

The *Memphis Commercial Appeal* warned the "colored soldier" to remember "his place." The article stated that regardless of past or future performance the black soldier should not think that "he was changed" or "benefitted his social condition by wearing a blue coat and carrying a gun."

These articles reflected the clashing attitudes in what was becoming an age-old controversy over the status of black men in uniform. Whites were not the only group to enter the debate. Even as the regular black regiments trained and prepared for embarkation to Cuba as professional soldiers, black civilian communities across the United States discussed the pros and cons of supporting the war. Those against argued that the United States would be hypocritical in sending black men into combat to free an oppressed people when a large segment of their own population still did not enjoy total freedom and equality.

Overall, the majority of African-American leaders encouraged black participation, as they had in every previous conflict. Booker T. Washington declared, "The Negro . . . will be no less patriotic at this time than in former periods of storm and stress."

The editor of the *Washington* (D.C.) *Colored American* agreed and wrote that white and blacks fighting together would unite the races. He concluded that such interaction would create "an era of good feeling the country over and cement the races into a more compact brotherhood through perfect unity of purpose and patriotic affinity."

Once the United States actually declared war against Spain, blacks as a whole united in support of the war effort. Individual black men volunteered in the same spirit of patriotism, adventure, and opportunity to prove themselves that has motivated soldiers of all races and causes since the beginning of time. In addition, they continued to believe, as had black men in every prior American war, that honorable service would decrease discrimination and improve their quality of life.

The *Chicago Tribune* expressed these thoughts in a June 1, 1898, editorial:

> If the colored men respond to the call for their services in carrying out this war . . . and prove themselves the brave soldiers they did towards the close of the Civil War and have done since, it is to be hoped that it will have the effect to remove the prejudice which

> has stripped them of their political rights in so many of the states and so often made them the victims of mob violence. If the bravery and patriotism manifested by colored men . . . shall accomplish this result, it . . . will be worth all the war will cost them and the nation at large.

And respond is exactly what black men of America did when President McKinley called for volunteers. They were ready to enlist, but there were initially few places for them to go. After the first request for volunteers, many states limited enlistments to whites. Only Alabama, which faced difficulties in meeting its quota, and Massachusetts and Ohio at first authorized black militia units. However, when the shortage of white manpower became apparent in other states, the governors of Illinois, Indiana, Kansas, North Carolina, and Virginia authorized black units after the president issued his second call for recruits.

The War Department planned to federalize these militia units and assign them to the Regular Army after their training. The plan quickly met resistance from some officers who did not wish to command black troops. When the War Department announced the assignment of the black Ninth Ohio Infantry Battalion to the U.S. Second Corps during the predeployment buildup stage, the corps commander, Brig. Gen. William C. Oates, strongly objected. Oates acknowledged that the black battalion exhibited adequate training and discipline but expressed his concern that being known as a commander of African Americans would taint his political future when he returned to his home state of Alabama. Instead of cashiering Oates and putting him on the next train south, the War Department transferred the Ninth Ohio to another command, leaving the general with only white units under his control.

In addition to federalizing state militias to augment the army, Congress authorized the formation of an additional ten infantry regiments, to be composed of volunteers from the Deep South who, they thought, would be resistant to tropical diseases, such as yellow fever. Four of these "immune regiments"—the Seventh, Eighth, Ninth, and Tenth Regiments of the USVI—were designated as black units.

Plans called for the four regiments and the black state militias to be commanded by white officers until black leaders insisted that black units be led by black officers. John Mitchell Jr., of the *Richmond Planet*, established the rallying cry of "No officers, no fight," and black leaders encouraged potential volunteers to postpone enlisting until the army agreed to their demand. Some state militias conceded by commissioning former Regular Army enlisted men and awarding the rank of officer to community leaders regardless of military experience. Ultimately, the entire chain of command of the Eighth Illinois, Twenty-third Kansas, and Third North Carolina Regiments was black.

These black officers, recognizing their exemplary positions, knew they must set the standards. The commander of the Eighth Illinois declared, "If we fail, the whole race will have to shoulder the burden."

The Regular Army also yielded, at least partially, to the demand for black officers. In the "immune regiments," blacks could become officers but could not advance beyond the rank of lieutenant. All senior commanders and staff officers above the grade of captain would be white. During the one-year existence of the USVI regiments, approximately one hundred black men secured commissions, about thirty of whom were Regular Army soldiers later reassigned as officers in the volunteer units as a reward for their outstanding combat performance in Cuba.

For all the recruiting, controversy, and effort, the volunteer units—army and militia—missed the war because of the brevity of the fighting. The only black state militia unit to actually engage the enemy was L Company of the Sixth Massachusetts, a black company in an otherwise white regiment. This action occurred during the brief and mostly bloodless invasion of Puerto Rico a week after the surrender of the Spanish in Cuba. Otherwise, the only African-American units to participate in actual combat in Cuba were the four Regular Army regiments, and they had to fight to do so every step of the way—against white Americans as well as the Spanish.

In early June the regiments, enduring jeers and curses along their rail routes, assembled at Tampa, Florida, for embarkation, only to find the port city segregated. Several fights occurred as a result of white soldiers harassing the blacks with name-calling. In a May 8 letter to his wife, the white commander of the Twenty-fourth Infantry, Col. Jacob Kent, wrote of Tampa, "This is not a nice town for my men. The feeling is strong against their color."

The most serious incident occurred during the evening of June 6 when members of the Twenty-fourth and Twenty-fifth Infantry became involved in a fight with an Ohio regiment whose members had been using a two-year-old black boy as target practice—seeing how close they could shoot at him without actually hitting him. Outraged, the black soldiers attacked the white "sharpshooters," and a fracas erupted. Local military authorities called in the Second Georgia Volunteer Infantry to restore order. The white infantrymen cleared the streets, seriously injuring about thirty of the black soldiers but only a few of the Ohio militiamen.

The incident played in the press with predictable perspectives. The black newspapers in the North condemned the assault on the black soldiers in Tampa, while the *Atlanta Constitution*, making no mention of the discipline level of the Ohio "marksmen," reported that the riot demonstrated that "army discipline has no effect on the Negro."

The discrimination against the black regiments continued when they boarded their transports to Cuba. For starters, limited ship space precluded the black Ninth and Tenth Cavalry regiments from taking their horses, which meant they would have to operate as foot soldiers once in Cuba. Ship captains then assigned the black regiments to the lowest decks and separated portions of the vessels into black and white areas. The final insult occurred when rumors of Spanish warships in the area postponed their sailing for a week. The black soldiers had to remain onboard, while whites could go ashore for bathing and recreation.

After finally getting clearance to sail, the invasion force approached the Cuban coast. On June 22, 1898, at the initial landing site near Daiquiri, heavy surf so battered the small boats taking the soldiers ashore that some capsized. Two Tenth cavalrymen drowned before the invasion fleet moved down the coast a few miles to Siboney and a better harbor that protected the landing of the remainder of the force.

On shore the Tenth Cavalry, along with the white First Cavalry Regiment and the First Volunteer Cavalry Regiment (Rough Riders), came under command of one of the few senior Confederate officers to be reinstated in the U.S. Army following the Civil War. Although Maj. Gen. Joseph Wheeler still tended, in the heat of battle, to refer to the opposing Spaniards as "yankees," he nevertheless proved a good leader. Wheeler's boss and overall commander of the ground operation, Maj. Gen. William Shafter, understood and appreciated the capabilities of the black soldiers from his experience with the Twenty-fourth Infantry in the 1870s.

Troop F, Tenth Cavalry, shortly after returning from Cuba (ca. 1902). (U.S. ARMY MILITARY HISTORY INSTITUTE)

After only one day of preparations, Wheeler's command moved inland from Siboney and late on June 23 became engaged at Las Guásimas in the first battle of the Spanish-American War. The battle began with the Rough Riders assaulting a heavily defended Spanish infantry position protected by trench networks and stone blockhouses. After the Rough Riders suffered several casualties, the two regular cavalry regiments joined the fight. Within the next ninety minutes the three regiments overran the Spanish stronghold; sixteen were killed, and another sixty were wounded.

Observers of the fight had great praise for all combatants, including members of the Tenth Cavalry. Stephen Bonsal, a war correspondent, later wrote in the August 1898 edition of the *Southern Workman* about the black soldiers: "They were no braver certainly than any other men in the line, but their better training enabled them to render more valuable services than the other troops engaged."

Word soon circulated throughout the army, especially among the black battalions in Cuba, that the Tenth Cavalry had saved the Rough Riders and their vocal executive officer, Theodore Roosevelt, from annihilation at Las Guásimas. Later in the war, and after its conclusion, when the stories of the Rough Riders began to dominate accounts of the conflict, black and white veterans alike often repeated the story of the rescue of the volunteer regiment. Some embellished it to the point of confusing the Battle of Las Guásimas with the later action on San Juan and Kettle Hills.

The best, and most accurate, account of what really occurred in the fight comes from the later writings of the Tenth Cavalry regimental quartermaster, Capt. John J. "Black Jack" Pershing. The captain gained his nickname from his service with African-American regiments. He would later command all U.S. troops in World War I. Observing the battle as a participant, Pershing noted that the Rough Riders would have suffered additional casualties without the timely arrival of the First and Tenth Cavalry Regiments. He directly credited the Buffalo Soldiers with "relieving the Rough Riders from volleys that were being poured into them from that portion of the Spanish line."

During the next week the Americans brought more regiments ashore and moved them forward in preparation for attacking the high ground above the port of Santiago de Cuba, where the Spanish fleet lay at anchor, blockaded from the open sea by the U.S. Navy. The terrain and limited road network dictated a rather simple plan of two direct attacks—one against the village of El Caney and another designed to capture San Juan and Kettle Hills.

On July 1 the Twenty-fifth Regiment joined six thousand white soldiers in the early-morning assault against El Caney, where they encountered five hundred to six hundred Spaniards, with their defense system that included several wooden blockhouses, a stone fort, and a trench network supported by barbed wire and rifle-firing pits. Advancing across an open field toward the Spanish defenses, the lead members of the Twenty-fifth and the white regiments began falling from enemy rifle fire. The attack slowed as the Americans sought cover in ravines and other low areas, and the best marksmen began sniping fire to pick off the defenders.

The sniping tactic yielded so many enemy casualties that the defenders, quickly demoralized, surrendered before the Americans had to make a final assault. Sharpshooters from the Twenty-fifth played an important role in breaking the Spanish defense, and the regiment's casualties of seven killed and twenty-five wounded matched that of the white regiments in the battle. Pvt. Conny Gray of the Regiment's D Company saved his white company commander's life by advancing under fire to pick up the wounded captain and carry him to medical care in the rear of the battle area.

The Battle of El Caney was still in progress when the American main force began its assault against San Juan and Kettle Hills. The Twenty-fourth Regiment anchored the extreme left flank of the attack, a section which quickly encountered withering rifle and artillery fire from the heights above. One of the few white volunteer regiments to join the fight, the Seventy-first New York, broke and ran, leaving the heaviest fighting to the Regular Army units. The black soldiers of the Twenty-fourth never wavered as they secured the flank and continued to advance. Six of the black infantrymen and one of their white officers fell to Spanish fire. Another fifty-six enlisted men and seven officers suffered wounds.

At the end of the battle, the Twenty-fourth Regiment stood on top of the hill amid dead and captive Spaniards. Capt. Benjamin W. Leavell, commander of A Company, praised his men for their spirit to "do their duty, yea, more than their duty." Leavell summed up the performance of the entire Twenty-fourth Regiment, stating, "Too much can not be said of their courage, willingness, and endurance."

On the far right flank of the attack against San Juan and Kettle Hills, the Ninth and Tenth Cavalry, both fighting dismounted as infantry, faced defenses similar to those encountered by the Twenty-fourth Infantry on the left side. As the Americans neared the Spanish defensive positions, regiments became intermixed, black and white soldiers fighting side by side. When the color-bearer of the white Third Cavalry fell from rifle fire, Sgt. George Barry, color-bearer of the Tenth Cavalry, grabbed the flag of the Third and carried both to the top of the hill. Never wavering, the black cavalrymen continued the assault until the Spanish, entrenched on the heights, surrendered. By the end of the fight, casualties in the Ninth and Tenth Cavalry Regiments totaled seven enlisted men and three officers killed and another eighty-one soldiers and eleven officers wounded.

After the Battle of San Juan Hill, "Black Jack" Pershing declared, "We officers of the Tenth Cavalry could have taken our black heroes into our arms. They had fought their way into our affections, as they have fought their way into the hearts of the American people."

All across the island, Cubans and white American soldiers echoed the praise of the performance of the black regiments. Even their Spanish adversaries respectfully referred to them as "Smoked Yankees." In the ensuing months, drawings and prints of the soldiers' assault up San Juan Hill appeared proudly in black publications and on the walls of African-American homes across the United States. The *Army and Navy Journal* reported, "Never in history has the Negro advanced himself so rapidly in public estimation as in this war."

Theodore Roosevelt referred on several occasions to the black soldiers as "brave men, worthy of respect." He also acknowledged: "I don't think any Rough Rider will ever forget the tie that binds us to the Ninth and Tenth Cavalry."

The victory at San Juan Hill brought most of the fighting on the Cuban mainland to an end. On July 3 the United States and Spain declared a truce; the Spanish formally surrendered on July 17, 1898.

The black regiments did not remain idle during the weeks of peace negotiations. They dug trenches, processed prisoners, and established camps. On July 15 the hospital at Siboney requested the Twenty-fourth Regiment to assist in the wards of soldiers suffering from yellow fever. Although eight white regiments had turned down the request, the Twenty-fourth willingly accepted and performed magnificently. Over the next month more than half the regiment came down with the disease. At the height of the epidemic, 241 of the Twenty-fourth's soldiers were themselves patients in the hospital, disproving the theory of black "immunity" to tropical diseases.

During the latter part of 1898 and in early 1899, the Ninth USVI and the Eighth Illinois and Twenty-third Kansas militias participated in the occupation of Cuba, but the other black volunteer and state militia units never left the United States. By mid-1899 all of the volunteer and militia units mobilized for the war were disbanded, their members discharged. Black officers in the units lost their commissions, and those from the Regular Army regiments reverted to their enlisted ranks in their original units or were discharged.

Black sailors as well as soldiers also played a significant role in the liberation of Cuba and shed some of the war's first blood. Of the 266 sailors killed aboard the *Maine* in Havana harbor, 22 were African Americans. Approximately two thousand black sailors served as crew members of U.S. Navy ships during the Spanish-American War. Chief

Gunner's Mate John Jordan, aboard Adm. George Dewey's flagship the USS *Olympia*, fired the first shot of the overwhelming victory over the Spanish fleet in Manila Bay.

Another black sailor, Fireman First Class Robert Penn, earned the Medal of Honor aboard the USS *Iowa*, off Santiago de Cuba, on July 20, 1898. When a gasket exploded and boiling water flooded the engine room, Penn controlled the fire by standing only inches above the hot water on a board thrown across a coal bucket. His brave and immediate action prevented the boiler from exploding and causing the loss of the entire ship.

During the last few weeks of August 1898 most of the U.S. warships returned to home waters. With its defeat of the Spanish fleets in the Caribbean and in the Philippines, the U.S. Navy had earned the reputation as one of the strongest sea forces in the world. U.S. Army units also began making their way home, with many of the militias and volunteer units soon to disband. The four black Regular Army regiments sailed to Montauk Point, New York, to recuperate and refit.

The media as well as military and political leaders continued to laud the accomplishments of the black regiments during their first few weeks back home. Renowned reporter and novelist Richard Harding Davis wrote of San Juan Hill: "The Negro soldiers established themselves as fighting men that morning."

On October 8, the Tenth Cavalry Regiment marched down Washington's Pennsylvania Avenue before cheering crowds and passed in review for President McKinley. The city of Philadelphia invited all four regiments to a parade, and when the War Department explained that funds were not available for transportation, the city's residents personally paid the transportation costs for four troops of the Tenth Cavalry. Other cities and towns across the country sponsored black-unit parades and hosted luncheons and picnics for the returning heroes.

Black communities hoped that recognition of African-American accomplishments in Cuba and the welcome-home festivities signified a new respect that would lead to less discrimination. Unfortunately, that would not be the case, for blacks' success in the military did not alter whites' attitudes about equality. Concerned about competition for jobs from black civilians, whites persisted in their racism, unaffected by the performance of African Americans in the Spanish-American War.

Whatever celebrity the black soldiers enjoyed as a result of the war would be extremely short-lived.

By the end of the year—just five short months after the war—whites were ready to forget the contributions of the black regiments. On December 27, 1898, the *Washington Post* published an article headlined "No Praise for Negro" by its correspondent in Virginia. According to the reporter, a "packed audience" at the Richmond Academy of Music did not take kindly to actor and Rough Rider veteran Mason Mitchell's salute "to the gallantry and bravery of the 10th Cavalry," writing that "from all parts of the building came cries of 'put him out' and 'stop him,' and hisses drowned the voice of the speaker. Mr. Mitchell rebuked his audience, but to no purpose. . . . The hisses were continued until Mitchell had to ring down the curtain and retire from the stage."

By the time portions of the Tenth Cavalry traveled through the South in January 1899 to their new station in San Antonio, Texas, feelings against the black soldiers had risen to such a hostile level that locals at Meridian, Mississippi, and Houston, Texas, fired on the train cars. When part of the Tenth Cavalry reported for duty in Huntsville, Alabama, rumors spread that the local townspeople would pay a bounty for every dead black cavalryman.

The Ninth Cavalry found conditions no better at Fort Ringold, Texas, where several hundred citizens of nearby Rio Grande City fired upon the post's walls. White commanders of the Twenty-fifth Infantry Regiment in El Paso, Texas, reported that their soldiers suffered at the hands of the civilian police, who failed to enforce the law impartially and frequently jailed black soldiers for drunkenness or disorderly conduct without justification.

Col. Chambers McKibbin, commander of the military department in which the Twenty-fifth was assigned, noted, "There is unquestionably a very strong prejudice throughout all of the old states against colored troops. . . . It is not because the colored soldier is disorderly—for as a rule, they behave better than white soldiers . . . but because they are soldiers."

Black state militiamen and members of the USVI also found the white population anxious that the former soldiers quickly remember "their place" after their discharge. Members of the Atlanta police force met the train carrying the Third North Carolina home in early 1899 and went from car to car clubbing the unarmed soldiers. In Nashville the Eighth U.S. Volunteers met a similar welcome from seventy-five policemen and two hundred citizens, who beat the soldiers with pistols and clubs.

One of the police officers bragged, "It was the best piece of work I ever witnessed. The way they went for the Negroes was inspiring. And if a darky even looked mad, it was enough for some policeman to bend his club double over his head."

While tragic, it was not surprising that black veterans faced racism that had changed little since the Civil War. Although the most deeply rooted animosity remained centered in the Deep South, whites throughout the country exhibited similar feelings. Perhaps the greatest blow to progress in African Americans gaining respect and equality as a result of the Spanish-American War came not from the proclaimed racist but rather from a man black soldiers thought was a friend.

Theodore Roosevelt's comments during the war about the blacks being "an excellent breed of Yankees" and his praise of their fighting abilities and discipline ceased when he returned to American political life. Prompted by a desire not to alienate white voters and a crusade to make the Rough Riders the absolute heroes of the war, Roosevelt began to downplay the performance of the black regiments and ultimately to challenge their bravery and loyalty.

In his book *The Rough Riders*, Roosevelt damned the black soldiers with faint praise, stating that although they performed their duties well, they were "peculiarly dependent upon their white officers." Roosevelt also included his story about personally drawing his sidearm during the assault on San Juan Hill to stop a number of black soldiers running toward the rear. Although one of the black soldiers involved in the incident challenged this claim, explaining that the group stopped by Roosevelt had been on their way to secure desperately needed ammunition and medical supplies, the former Rough Rider made no retraction.

Roosevelt's comments, supported by other white officers who judged the black soldiers by the color of their skin rather than their performance under fire, reinforced the whites' belief that black soldiers performed well only when led by white officers. Roosevelt and his cohorts ignored the fact that many black noncommissioned officers (NCOs) displayed brave and sound leadership throughout the campaign.

The service of black women in the Spanish-American War received even less attention than the service of black men during and after the conflict. Although black and white women had served as nurses to the wounded of both sides during the Civil War, the U.S. military had not accepted women of any race in any capacity. The relatively small numbers of casualties during the Indians Wars, combined with the remoteness of thc western garrisons, influenced the army to keep its support forces, including medical, all male.

When the army mobilized for the Spanish-American War, the Daughters of the American Revolution (DAR) offered to recruit volunteer nurses and verify their qualifications so that the army could hire them on a contract basis to provide medical care for wounded soldiers returning from Cuba and for the larger numbers returning with tropical diseases. During 1898 the DAR provided fifteen thousand white women that the army hired as contract nurses.

While the DAR did not recruit or accept black volunteers, some government leaders, who believed that African Americans had a natural immunity to tropical diseases, encouraged their employment as contract nurses. Although their disease resistance proved just as much in error as that of the "immune" regiments recruited to fight in Cuba, the perception of their resistance opened the door for a few black women to serve.

In July 1898 the U.S. surgeon general authorized Mrs. Namahyoke Sockum Curtis, wife of the chief of surgery at Freedman's Hospital in Washington, D.C., to screen black nurses for contract employment by the army. Because the first school to train black nurses in the United States had not been established until 1886, the number of applicants was limited. Eventually, the army hired thirty-two black nurses, or less than three-tenths of 1 percent of the total number of women employed, to treat patients at posts where blacks departed and returned. The majority were assigned to Camp Thomas at Chickamauga Park, Georgia. All of these nurses wore civilian nursing attire rather then military uniforms.

The contract nurses, black and white, were discharged during the year that followed the Spanish-American War. The program had proved so successful, however, that Congress authorized an Army Nurse Corps in 1901 and a Navy Nurse Corps in 1908. For the first time, white women were a part of the regular armed forces, but black women remained excluded.

Less than a year after the battles of Las Guásimas, El Caney, and San Juan and Kettle Hills, the accomplishments of the Ninth and Tenth Cavalry and the Twenty-fourth and Twenty-fifth Infantry Regiments became nothing more than a downplayed footnote in history, as recorded by white academics.

In reality, of course, black regiments were neither a footnote nor historical artifacts. They were alive and well and members of the U.S. Army about to face yet another call to duty. The four black Regular Army regiments barely had time to settle into their new assignments along the U.S. border with Mexico before receiving orders to report to San Francisco for transfer to the Philippines. George Dewey's fleet, having originally attacked Spanish ships in Manila Bay and secured the Philippines as a bargaining chip in peace negotiations, had unwittingly acquired for the United States a long-term trouble spot instead of a valuable prize. Filipino rebels, pleased with the American defeat of the Spanish, were now demanding independence rather than occupation.

On February 4, 1899, Filipino rebels under the command of Emilio Aguinaldo began combat operations against their "American occupiers." Two days later, the U.S. Congress reacted by authorizing the annexation of the Philippine Islands. On March 2, Congress approved a call-up of thirty-five thousand volunteers to reinforce the Regular Army's campaign against the rebels.

The buildup sparked the same concerns as had preparations for the Spanish-American War. Many black leaders, including Booker T. Washington, complained about the treatment of black veterans in the United States and argued that the government should strive for equality at home before imposing American "democracy" on foreign soil. Other black leaders, however, echoed the same pre–Spanish-American War sentiment: that African-American participation in U.S. military conflicts would combat prejudice and lead to better overall treatment.

Both black and white leaders expressed concern about African-American soldiers fighting a dark-skinned enemy. Most agreed that the discipline and past performance of the four black Regular Army regiments guaranteed their loyal service, but some officers posed questions about the loyalty of volunteer regiments. As a result, the army formed only two additional regiments of blacks, the Forty-eighth and Forty-ninth Volunteer Infantry, for the conflict.

A total of seventy-five black soldiers—volunteers reactivated from the Spanish-American War units and NCOs from the Regular Army—received commissions in the two volunteer regiments. In contrast to the previous war, when they were only allowed to serve as lieutenants, black officers could now advance to the grade of captain, though the field-grade ranks remained the domain of whites, ensuring that only they occupied the regimental command and principal staff positions. Again, as in the Spanish-American War, none of the commissions in the volunteer regiments were permanent appointments; they would expire at the end of the conflict or the unit's disbandment.

While the volunteer regiments assembled and trained, the regular regiments prepared to embark to the war zone. On June 19, 1899, the *San Francisco Chronicle* published an article expressing doubt as to whether black soldiers would fight their dark-skinned "brothers" in the Philippines. Members of the Twenty-fourth Infantry Regiment, waiting at the San Francisco docks for their transports, angrily responded to a *Chronicle* reporter that the article was "a libel on the regiment." In an interview, a member of the Twenty-fourth summarized the general feelings of black soldiers, "We are American citizens," the soldier proudly declared, "and we have at heart the interests of our native land in the same manner as do all Americans."

On July 2, after the Twenty-fourth had already sailed westward, the *Chronicle* reversed itself and published a long tribute to the black soldiers. The article asked all Americans to remember: "This is a man and a brother, no longer the bond slave but a citizen who of his own free will gives of heart and brain and brawn to the cause of the republic."

By the end of the summer of 1899 all four regular black regiments had arrived in the Philippines. Early in 1900 the two black volunteer regiments joined other U.S. forces in what had become mostly a counterguerrilla war in the islands' jungles. The rebels immediately targeted the black soldiers with propaganda, asking how they could fight to oppress the Filipino people when blacks faced discrimination and lynching in the United States. To further exploit racial tension among their enemies, the rebels promised commissions in their army to black deserters, but only five black Americans crossed the lines during the entire campaign.

The most notorious of these men was David Fagan, who deserted in November 1899 from I Company, Twenty-fourth Infantry, and advanced to the rank of captain in the rebel army. Motivated to desert more by difficulty in adapting to his regiment's discipline and following his sergeant's orders than from any political ideology, Fagan fought alongside the rebels until a Filipino bounty hunter killed him in December 1901.

Fagan's desertion had far greater consequences for blacks than just his combat action against his former comrades. Theodore Roosevelt, who became president after McKinley's assassination, commuted the death sentences of fifteen white soldiers convicted of desertion, but he did not stop the execution of two black soldiers accused of the same offense. Although more whites deserted in the Philippines than did blacks, Roosevelt ordered the execution of Pvts. Edmond DuBose and Louis Russel, both of the Ninth Cavalry Regiment, to deter further desertions.

Despite the propaganda by the rebels, the overall performance by the black regiments in the Philippines differed little from their service in previous wars. They fought bravely and loyally alongside white regiments, with whom they shared the dangers of the battlefield, the threat of disease, and the loneliness of being far from home. Most of the black soldiers' sentiments about fighting the Filipino rebels closely resembled those of M. W. Saddler of the Twenty-fifth Infantry, who wrote shortly after the regiment's arrival in September 1899, "We are now arrayed to meet a common foe, men of our own hue and color. Whether it is right to reduce these people to submission is not a question for a soldier to decide. Our oaths of allegiance know neither race, color, nor nation."

The treatment of black regiments in the Philippines also differed little from previous wars. The same Jim Crow policies of the American South followed the soldiers to the Philippines, where blacks found themselves excluded from "whites only" eating establishments, barbershops, and other public facilities. Black officers found little acceptance from their white counterparts or respect from white enlisted men. Those who refused to salute the rank of black officers received no rebuke nor punishment from white commanders.

Despite the lack of equality, the black regiments remained professionals. Most of their operations in the Philippines, like those of the white regiments, focused on pursuing guerrillas and searching for their villages and retreats. The actions surrounding San Augustin on October 7, 1899, were typical. Members of the Twenty-fourth Infantry waded waist-deep through swamps to approach the enemy camp, only to have the majority of the guerrillas disappear into the jungle after the first rifle volley. It then took the infantrymen several more days of pursuit before they flushed out the majority of the rebel band.

Other black soldiers in the Twenty-fourth Regiment participated in one of the most successful operations of the Philippine insurrection. Soldiers of H and K Companies, supported by detachments of A and K Companies, were under the command of Capt. Joseph B. Batchelor. They departed Cabanatuan on November 23, 1899, and advanced through remote mountains in central Luzon to attack a rebel base area. According to Batchelor's official report, they marched more than three hundred miles over unmapped territory without guides, using trails that were "just passable," enduring "chilling nights and sweltering days," while crossing "123 deep fords" and living for three weeks "on unaccustomed and insufficient food."

On December 7 the trail-weary command of 350 men forced the surrender of rebel general Daniel Tirona and his thousand-man army. Batchelor concluded his report by modestly adding that the opposing soldiers and the Filipino civilians in the pacified area were now "enthusiastic advocates of American supremacy."

Gen. Elwell S. Otis, one of Batchelor's superiors, lauded the success of the Twenty-fourth Regiment in an official report, noting that the operation was "memorable on account of the celerity of its execution, the difficulties encountered, and the discomfort suffered by the troops."

The continuing success of the U.S. military units, such as the Twenty-fourth's capture of Tirona and his army, took its toll on the rebel efforts. Progressively, the number of combat engagements disintegrated into sporadic skirmishes. The final blow to the concerted effectiveness of the Filipino insurgents was the capture of its leader, Aguinaldo, on February 4, 1901.

With the rebellion reduced to cleanup operations, the army began its transfer of units back to the States. The black volunteer Forty-eighth and Forty-ninth Infantry Regiments returned home in 1901 and were disbanded, its members either discharged or accepted as enlistees in the Regular Army black regiments. None of the officers could retain their commissions. Their choices were to accept discharge or revert to the enlisted ranks.

In 1902 the black Regular Army regiments also rotated back to America to take up their posts along the U.S.-Mexican border. Several units from the regiments returned to the Philippines to serve brief occupation-duty tours and to combat the final guerrilla holdouts. By 1906 all were back in the States.

Despite their continued exemplary performance in combat, black soldiers and their counterparts in the navy faced persistent racism and discrimination. Loyal service in war produced few peacetime benefits for African Americans wearing uniforms of the U.S. military. The next decade would again demonstrate that while they reluctantly accepted blacks in uniform in time of war, the majority of white Americans had little tolerance for black soldiers and sailors in time of peace.

6

THE BATTLES OF PEACE: BROWNSVILLE AND THE GREAT WHITE FLEET

During the first decade and a half of the twentieth century, the plight of black servicemen significantly worsened despite stellar performances on both land and sea through five wars. Indeed, the early 1900s found African Americans in the military facing a bleak future, with their very right to serve threatened and their opportunities limited.

From its very beginnings, the U.S. military had mirrored the civilian society. Never had that been more true than in the early part of the twentieth century, when the military reflected civilian efforts to increase segregation and the separation of the races. All across the United States a growing population of white landowners and merchants began increasing their efforts to block black Americans from any position of authority. These whites, along with increased competition with European immigrants for jobs, seemed determined to isolate all African Americans except servants and laborers from the rest of the population.

While the overall number of army regiments and navy ships increased in the postwar period, the number of black soldiers and sailors on active duty not only did not keep pace with the expansion but also declined to levels below those of the previous decade. On February 2, 1901, Congress authorized an additional 1,135 officers for the

Regular Army. Most of them came from the volunteer regiments and received their regular commissions on the recommendations of their commanders. Not a single black received one of those commissions.

The Colored American Magazine published an editorial on June 6, 1903, that both asked and answered the question in the minds of many African Americans about the lack of commissions for blacks: "What does this mean? Does it mean that the brave black soldiers who volunteered to stand by the flag are to be ignored in the reorganized army? . . . It would seem so."

Separation between black and white enlisted men escalated as well, whites emphasizing that one should stay with one's "own kind." A white soldier in New Jersey discovered to what extent racial lines had been drawn when he married a black woman in 1904. The army quickly discharged the white soldier on the grounds that his marriage made his "future retention in the service prejudicial to the public interest." Seven years later a black soldier in the Twenty-fifth Infantry received a discharge after marrying a white woman because the marriage made him "unfit for further association with other soldiers."

In addition to the authorization for more officers in 1901, Congress also directed the formation of five more infantry and five more cavalry regiments. The ten regiments were to be all white.

Blacks also realized fewer opportunities to serve in the state militias—now known as the National Guard. Both the military and the state governments defined the Militia Act of 1903 as limiting federal control of the National Guard. All agreed that the racial composition of the National Guard remained a state prerogative, and as a result, many state units no longer accepted blacks. By the time the United States entered World War I in 1917, the National Guard contained only five thousand African Americans, less than 3 percent of its total, and most of them served in the Eighth Illinois—originally organized with black volunteers for the Spanish-American War—and the newly formed Fifteenth New York. Single black companies of about one hundred men each represented Massachusetts, Maryland, Ohio, Tennessee, and the District of Colombia. No black unit or black man served in the National Guard of any of the Deep South states.

Complaints about the dearth of black officers and the paucity of black Regular Army and reserve units did produce a few "token" changes. On February 8, 1901, the army appointed John R. Lynch, an officer in the volunteers, as the first black Regular Army captain in the Paymaster Department. Benjamin O. Davis, an enlisted man in I Troop, Ninth Cavalry, received a Regular Army commission as a cavalry second lieutenant on May 19, 1901. Davis, a former first lieutenant with the Eighth Volunteer Infantry during the Spanish-American War, had reverted to his enlisted rank after the conflict ended. He became the first black soldier to rise from the ranks to become a Regular Army officer. Another black enlisted man, Cpl. John E. Green, of the Twenty-fourth Infantry, received a Regular Army commission the following June.

Both Davis and Green earned their commissions through a series of examinations in competition with other black enlisted men. Although more than a dozen qualified, only Davis and Green received commissions. The total number of combat-arms black officers in the Regular Army regiments still numbered fewer than the fingers on one hand. Except for a small number of black chaplains, the vast majority of black soldiers still served under an all-white chain of command.

The token appointments did little to appease black activists, who continued to lobby for more black officers and additional black units. They also continued to campaign to open the artillery branch to African Americans. Both the majority of congressmen and senior army officers stood by the Civil War–vintage argument that black soldiers lacked the intelligence to master the technical skills required of artillery crewmen.

During the late summer of 1906 an incident occurred in Brownsville, Texas, that reconfirmed for many white Americans the danger they perceived in blacks serving in the military. In the spring of 1906 the War Department ordered black soldiers of Companies B, C, and D of the First Battalion, Twenty-fifth Infantry Regiment, from Fort Niobrara, Nebraska to Fort Brown, a post at the mouth of the Rio Grande River, just outside the town of Brownsville.

In issuing the orders, the War Department further directed the First Battalion—minus A Company, which was assigned to Wyoming and scheduled to join the rest of the unit later in the year—to stop en route to conduct joint training with the Texas National Guard at Camp Mabry, near Austin. Both the black soldiers and the white officers of the Twenty-fifth Regiment protested the orders, citing previous trouble with the racist Texas guardsmen on maneuvers at Fort Riley,

Kansas, in 1903 and the general treatment by Texans during the regiment's previous assignments to the state. Theophilus G. Steward, the regiment's chaplain and one of the few black officers on active duty, summed up the feelings of the unit, stating, "Texas, I fear, means a quasi-battleground for the Twenty-fifth Infantry."

The citizens of Brownsville also protested the reassignment, and the local U.S. commissioner wrote Secretary of War William Howard Taft requesting that he revoke the orders for the Twenty-fifth and instead keep the white Twenty-sixth Infantry at Fort Brown. After much debate, the War Department canceled the Camp Mabry duties but refused to repeal the order that the First Battalion report to Texas. Taft responded to the Brownsville commissioner:

> The fact is that a certain amount of race prejudice between white and black seems to have become almost universal throughout the country, and no matter where colored troops are sent there are always some who make objections to their coming. It is a fact, however, as shown by our records, that colored troops are quite as well disciplined and behaved as the average of other troops, and it does not seem logical to anticipate any greater trouble from them than from the rest.

The First Battalion arrived at Fort Brown on July 28, 1906, without incident. Although saloons, restaurants, and other public facilities had posted Jim Crow laws barring blacks from entrance and the local park now sported a new sign announcing "No Niggers or Dogs Allowed," the black companies settled in and took their recreation at a bar which two African-American ex-soldiers had opened to serve them.

The calm surrounding the arrival of the Twenty-fifth lasted less than two weeks, and when the trouble began, it set off a chain of events marked by injustice, abuse, humiliation, and devastation for the black soldiers unfortunate enough to find themselves in the Rio Grande Valley.

On August 5, two black soldiers, Pvts. James W. Newton and Frank J. Lipscomb, encountered Fred Tate, a local white customs inspector as he escorted several white women down a Brownsville street. According to Tate, the soldiers jostled the women, provoking him to draw his revolver and strike Newton on the head, knocking him unconscious. Newton and Lipscomb, described by their white commanding officer as "very reliable" and "very inoffensive in their manner," claimed that they had done nothing to provoke Tate, whom they had overheard say just before drawing his gun, "I'll learn you to get off the sidewalk when there is a party of ladies on the walk."

A week later, on August 12, A. Y. Baker, another white customs official, threw Pvt. Oscar W. Reed of C Company from a bridge into the Rio Grande River. Baker reported that he had taken the action because Reed had drunkenly argued about the crossing fee. Later that same afternoon, a local woman claimed that a black man had attempted to molest her and had fled when she screamed. A headline in the *Brownsville Daily Herald* the following day declared, "Infamous Outrage—Negro Soldier Invaded Private Premises Last Night and Attempted to Seize a White Lady."

That afternoon, Brownsville mayor Frederick J. Combe met with First Battalion and Fort Brown commander Maj. Charles W. Penrose. Both men, fearing reprisals by the Brownsville citizens against the soldiers, agreed that the infantrymen adhere to a post curfew of 8:00 P.M. until tempers cooled. Some of the soldiers had already departed on pass, so Penrose sent patrols into town to return all of his men to the fort. By 10:00 P.M. company commanders accounted for all but three of the battalion's soldiers.

A few minutes after midnight on the morning of August 14, a group of men numbering between six and twenty gathered across the road from Fort Brown, opposite the B Company barracks, at a narrow passage known as Cowen's Alley. The men then began firing shots into buildings and at streetlights and proceeded to shoot up the alley for about ten minutes.

The random bullets killed bartender Frank Natus; seriously wounded police lieutenant M. Y. Dominguez, who would have to have his arm amputated; and slightly wounded Paulino S. Preciado, the editor of a local Spanish newspaper.

At the sound of the first shots, the sergeant of the guard alerted all of Fort Brown to stand ready for attack, initially assuming that the post was under fire by the townspeople. Even as the bullets were still being fired, companies took roll calls, and each commander reported all present, including the three unaccounted for at 10:00 P.M.

Mayor Combe and Major Penrose met during the night to determine the identity of the raiders, because several Brownsville citizens claimed to have seen black men among the shooters. At first light, Penrose led an inspection of the battalion's arms racks and found all weapons clean and with no evidence that they had been fired. Local civilians began coming forward, however, with military rifle cartridges and clips from Springfield rifles recently issued to the Twenty-fifth Regiment, supposedly the only source of such ammunition.

Combe organized an investigative committee of prominent local citizens who interviewed twenty-two civilian witnesses, none of whom were under oath. Only five of the witnesses claimed to have seen black soldiers in the streets at the time of the shootings; another individual professed to have only heard voices that sounded like blacks talking. The committee made no effort to call any soldiers before their inquiry.

Questions to the witnesses revealed the committee's purpose and predisposed ideas about the raid. When local citizen Herbert Elkins appeared before the committee, they prefaced their questioning by stating: "You know the object of this meeting. We know that this outrage was committed by Negro soldiers. We want any information that will lead to a discovery of who did it."

The committee greeted another witness thusly: "We are inquiring into the matter of last night with a view to ascertaining who the guilty parties are. We know they were Negro soldiers. If there is anything that would throw any light on the subject, we would like to have it."

After less than two days of testimony, the committee, satisfied that they had confirmed their original beliefs about the guilt of the black soldiers, telegraphed President Theodore Roosevelt on August 15: "Our condition, Mr. President, is this: Our women and children are terrified and our men are practically under constant alarm and watchfulness. No community can stand this strain for more than a few days. We look to you for relief; we ask you to have the troops at once removed from Fort Brown and replaced by white soldiers."

On August 16, President Roosevelt ordered an immediate investigation and dispatched Maj. August P. Blocksom, assistant inspector general of the army's Southwestern Division, to Brownsville to take charge of the proceedings. Roosevelt also ordered the First Battalion to remain at Fort Brown until Blocksom completed his investigation. However, renewed protest by the people of Brownsville changed his mind. On August 25 all but a dozen members of the battalion were transferred by train to Fort Reno, Oklahoma Territory.

The dozen black soldiers who remained in Texas were prisoners at Fort Sam Houston in San Antonio, awaiting trial as members of the raid. Charges against these men were the result of a brief investigation by Texas Ranger captain William McDonald, who selected them as the most likely to be guilty, although he actually had no proof to support his accusations.

In a later official report of the Brownsville incident, a senior army officer noted, "The manner by which their names were procured is a mystery. As far as is known, there is no evidence that the majority of them were in any way directly connected with the affair. It seems to have been a dragnet proceeding."

A Cameron County, Texas, grand jury, meeting later in Brownsville, heard no convincing evidence despite much emotional testimony about the guilt of the dozen men. The court released the twelve soldiers, and they joined their regiment in Oklahoma.

Although neither the civilian nor military officials could determine just who fired the shots in Brownsville, both remained convinced that the guilty parties were among the black soldiers of the Twenty-fifth Regiment. Major Blocksom relied on the testimony of the white eyewitnesses in Brownsville and the shell casings and clips that apparently could have come only from the soldiers' Springfield rifles. He also placed the blame on the black unit.

In his August 29 report Blocksom stated, "That the raiders were soldiers of the 25th Infantry can not be doubted." Yet Blocksom admitted that neither he nor the regiment's officers could positively identify a single soldier by name, an impasse that prompted him to conclude that a "conspiracy of silence" existed among the battalion's enlisted men, including the black noncommissioned officers (NCOs), who surely must have known what occurred on the night of August 13.

Blocksom began the recommendations portion of his report by declaring that the First Battalion "had an excellent reputation up to the 13th of August, but the stain now upon it is the worst I have ever seen in the army." He then recommended that if the guilty parties did not come forward, "all enlisted men of the three companies present on the night of August 13 be discharged from the service and debarred from reenlistment in the army, navy, and marine corps."

Blocksom ended his report with an observation that revealed his own bias about black soldiers. The major noted, "It must be confessed that the colored soldier is much more aggressive in his attitude on the social equality question than he used to be."

"Social equality," not to mention common justice, ultimately eluded the members of the First Battalion. The most damning evidence against the soldiers continued to be the spent shell casings and clips. Army investigators discovered that the black companies had, upon receiving the new rifles back in Nebraska, fired the weapons, picked up the spent casings and clips from the ranges, and moved them to Fort Brown as part of the logistics process for eventual recycling. They had stored the shells and clips at unsecured areas within Fort Brown, awaiting shipment to a logistic center. Markings on the "evidence," including double firing-pin strikes and extraction marks, resembled those on casings from new rifles not completely clean—exactly like the ones ejected from the freshly issued, heavily lubricated rifles fired in Nebraska rather than from inspection-ready rifles in the arms room at Fort Brown. Information also surfaced that white soldiers who had preceded the Twenty-fifth Regiment at the post may have given or sold military ammunition and clips from their Springfields to the local civilians.

If the clips and casings were the damning evidence, the real force behind the investigation continued to be the people of Brownsville, who had objected to the assignment of black soldiers at the post before their arrival and who remained determined to have the black troops removed.

On October 4, President Roosevelt ordered still another investigation into the Brownsville incident, this time sending Brig. Gen. Ernest A. Garlington, the inspector general of the army and a native of South Carolina, to once again conduct interviews and study the evidence. Garlington reported on October 22 that he had interviewed the First Battalion soldiers and that each time he had asked about the incident, "the countenance of the individual being interviewed assumed a wooden, stolid look, and each man positively denied any knowledge in the affair."

Garlington admitted he could find no proof of collusion on the part of the members of the battalion to withhold information, but that fact did not deter his concurrence with Major Blocksom's conspiracy-of-silence theory. Garlington concluded: "The secretive nature of the race, where crimes charged to members of their color are made, is well known. . . . It has been established by careful investigation, beyond reasonable doubt, that the firing into the houses of the citizens of Brownsville . . . was done by enlisted men of the 25th Infantry. . . ."

Garlington endorsed Blocksom's recommendation to immediately dismiss all members of B, C, and D Companies of the First Battalion from the army. President Roosevelt approved the discharges but ordered that the announcement be withheld until after the congressional elections, held on November 9, "for fear of its effect on the colored vote."

Between November 16 and November 26, the army discharged all 167 soldiers of the First Battalion who were present for duty on the night of August 13 at Fort Brown. The army dispensed the discharges without honor, denied the soldiers all back pay and pension benefits, and barred them from reenlistment in any branch of the U.S. armed forces. None of the soldiers physically resisted their discharge, but several openly wept as they turned in their uniforms and other equipment. One of those discharged had loyally served for more than twenty-seven years; twenty-five had accrued more than ten years of active duty; fully half had been in uniform for more than five years; and more than fifteen had earned medals or certificates of merit for combat in Cuba or the Philippines.

The citizens of Brownsville as well as whites all across the country welcomed Roosevelt's decision. "Whatever may be the value of the Negro troops in time of war, the fact remains they are a curse to the country in time of peace," reported the *New Orleans Picayune*. Major media in the northern states also praised the president's action, with the *New York Times* writing that the black soldiers had only themselves to blame.

Blacks reacted with outrage. Booker T. Washington summed up the mood of blacks in general: "The race feels . . . hurt and disappointed." The *Crisis*, a black periodical, pointed out the discrepancies between the Brownsville episode and a recent incident in Ohio in which white soldiers had killed two civilians while attempting to break comrades from their unit out of a local jail. In that case, reported the *Crisis*, the white soldiers received a formal trial attended by legal counsel—a privilege denied the soldiers of the Twenty-fifth Regiment.

Of all the ink spilled and rhetoric expelled, the best summary of the discharge of the First Battalion came from the *Army and Navy Journal*. The *Journal* recognized the military implications as well as the legal impact and labeled the punishment a "drastic and arbitrary exercise of authority." The article also noted that the discharges took place on the recommendations of a southern-born officer. In its November 24, 1906, edition, the paper declared: "The finding against the Negro soldiers is based upon the testimony of white men given under circumstances that deprive it of all value as legal evidence." In still another article, the *Journal* stated that the president and the military had treated the soldiers of the Twenty-fifth Regiment like "a lot of 'plantation' niggers instead of as soldiers wearing the uniform of the army."

For many years after the discharge of the 167 soldiers of the First Battalion, the Brownsville incident remained a political issue. Debates, charges, countercharges, and presentations of evidence of doubtful veracity continued. None of this activity, however, revealed the complete truth of what occurred on August 13, 1906.

Politics superseded justice as both Roosevelt and Sen. Joseph B. Foraker of Ohio, a political opponent of the president and a fellow candidate for the Republican presidential nomination, used Brownsville as a rallying cry to further their own careers. On April 14, 1908, Foraker proposed a bill providing the men of the Twenty-fifth Regiment the opportunity to reenlist, proclaiming, "They ask for no favors because they are Negroes, but only for justice because they are men."

In one angry debate with Foraker, Roosevelt declared that at least "some" of the soldiers discharged were "bloody butchers" who deserved to be hung and whose lives were spared only "because I couldn't find out which ones did the shooting."

President Roosevelt attempted to provide a precedent for his discharge of the men of the First Battalion by citing two Civil War incidents in which Gen. Robert E. Lee ordered the mass dismissal of a regiment for poor overall performance in one incident and the discharge of another regiment for cowardice. Neither of these actions were substantiated, and to date the punishment of the black soldiers of the Twenty-fifth Infantry Regiment remains the only example of mass punishment without the benefit of a trial in U.S. military history.

Nonetheless, Foraker finally forced the reopening of the investigation, and in late November 1908 a board of officers announced that 14 of the 167 discharged soldiers would be eligible to reenlist. Eleven of the fourteen did ultimately reenter the army, though many believed that the army's accepting them was "tokenism," similar to the preincident promotion of a few black officers. The board offered no rationale for the exclusion of the other 153 soldiers who were not allowed to reenlist.

The token allowance of reenlistments and the passage of time dimmed the memories and animosity of Brownsville among the white population. In March 1971, sixty-five years after that explosive night in South Texas, Augustus F. Hawkins, a black congressman from California, introduced legislation to declare all of the remaining discharges "honorable." With the support of President Richard Nixon, the army reopened the investigation and in September 1972 announced that the records of the remaining 153 men of the Twenty-fifth Regiment would be amended to "honorable discharge." On December 6, President Nixon signed a bill authorizing a one-time pension payment of twenty-five thousand dollars to Dorsie Willis, eighty-six, the only survivor among those discharged.

What really happened at Brownsville will probably never be known. One thing is certain: that all the enlisted members of the First Battalion were not, and could not have been, involved. The soldiers of the Twenty-fifth Regiment had faced far worse treatment by white American civilians over the past decade, and their service had remained disciplined and professional. No evidence substantial enough to try even one individual soldier ever surfaced.

The real impact of the Brownsville incident, however, reached further than the dismissal of 167 soldiers. Brownsville was a significant turning point in the relationship between blacks and whites in the army. White officers who believed in the guilt of the First Battalion lost confidence in the discipline and capabilities of black soldiers. More importantly, African-American soldiers lost confidence in their officers' abilities and willingness to properly represent them and in the military's promises of equal justice for equal performance. For nearly a half century during and after the Civil War black soldiers had striven to prove themselves as professional soldiers and their race as deserving of equality both in and out of uniform. Brownsville proved to be pivotal in diminishing the gains for which they had fought so hard.

Despite the loss of confidence between black and whites in the army, African-American leaders continued their efforts to expand the presence of blacks in the military services. Additional requests to open the artillery branch to blacks finally led to a U.S. Army War College study at Carlisle Barracks, Pennsylvania, that began in 1906. The War College report recommended the continued exclusion of African Americans from the artillery because of the technical skills required to properly operate the guns and produce accurate, safe fire. In summary, the report declared that blacks overall were "inferior to the white race in intelligence and mental abilities."

The War College findings echoed the pre–Civil War studies that had justified slavery on the basis of blacks being an inferior species to whites and suitable only for menial, subservient positions. Many whites, regardless of social class or station, could not conceive of African Americans possessing equal intelligence or being worthy of equal treatment and opportunity. Armed black soldiers threatened their perception of personal safety as well as their concept of the "Negro's place," and that "place" certainly was not achieving the rank of officer or serving in any branch of the service considered prestigious.

The Brownsville incident and subsequent investigations did nothing to counter the War College report. The politically minded secretary of war, William Howard Taft, used the report to deny the entry of blacks into the artillery branch. In an attempt to keep all sides happy, Taft recommended a token authorization of black bandmasters. As a result, in November 1908, President Roosevelt ordered that all colored regiments be authorized black bandmasters to replace the previously white occupants of those positions.

Instead of an artillery regiment of five hundred black soldiers or even an artillery battery of a hundred men, two years of studies and lobbying produced a gain of a grand total of four additional black bandmaster positions. Tokenism had reached a new low.

The U.S. Navy during this period suffered no racial incidents of the magnitude of Brownsville, but black sailors experienced the same increase in prejudice and discrimination felt by African-American soldiers. As long as U.S. Navy ships remained sailing vessels, black and whites coexisted amiably enough, because all crew member experienced the same hardships and hazards of life at sea. However, as the navy moved from sail to mechanical power and living conditions onboard began to improve, prejudices surfaced and flourished.

Between 1894 to 1906 several brawls occurred between black and white sailors aboard U.S. Navy ships, including the *Charleston*, *Columbia*, *Boston*, and *Independence*. In a letter to the editor of the *Army and Navy Journal*, published on January 5, 1907, Lt. George Steunenberg wrote that integrated crews did not function well on his last ship. According to Steunenberg, "the presence of blacks was a constant source of dissatisfaction which often broke out in bloody fight."

In addition to dealing with increased racial tensions, black sailors at the beginning of the twentieth century also faced fewer enlistment opportunities and further restrictions on the duties they could perform as well as the rank they could achieve. As its ships became fully steam powered, the navy changed its recruiting tactics. Recruiters no longer sought seamen experienced in climbing masts and unfurling yards of sails. Nor did they limit their recruitment to East Coast seaports from which they had traditionally signed up both whites and blacks. Naval recruiters now opened offices across the United States and narrowed their selection process.

Whereas recruiters had historically accepted knowledgeable sailors regardless of their past criminal records, abuse of alcohol, or skin color, they now could selectively enlist nationwide. Navy recruiting stations in the Midwest and elsewhere signed on bright, well-behaved young white men fully capable of learning how to operate steam-powered, steel-hulled ships. Even though the navy significantly increased its number of ships and these larger, more modern vessels required more crewmen, recruiters managed to easily fill their quotas. As more and more whites came forward to join the "new navy," recruiters, of their own volition, began to accept fewer black volunteers. In 1901 the Navy Department reinforced the recruiters' bias by informing them that they were no longer to seek out African-American recruits.

Those blacks already in the navy and the few new recruits allowed to enlist found fewer jobs. They could work only in the galleys as servants to officers or in the boiler rooms as firemen shoveling coal. These restrictions were soon to become a matter of official policy.

Oddly enough, it was white prejudice against another minority race that set the course for the immediate future of blacks in the navy. Japanese Americans on the West Coast suffered discrimination similar to that of blacks in the civilian community, and Japanese sailors served in the navy almost exclusively in the messman's branch, a formalized specialty of galley workers and servants in the officers' dining rooms established on April 1, 1893.

But unlike blacks, Japanese Americans had the support of their motherland, which was both a blessing and a curse. After defeating Russia in 1905 and becoming a world power in contention with the United States and other countries for control of the Pacific Ocean, Japan began expressing concern about the poor treatment of their countrymen in America. In response to Japan's growing influence, President Roosevelt planned to dispatch a fleet of sixteen battleships—all built since the end of the Spanish-American War, including the new USS *Maine*—on a round-the-world cruise as a show of power. The ships, all painted white to be even more imposing, immediately became known as the Great White Fleet.

While the Great White Fleet assembled at Hampton Roads, Virginia, for its voyage, naval leaders grew concerned about the number of Japanese messmen aboard the battleships. Some feared espionage or sabotage by the Japanese sailors, while others worried about foreign reaction to Oriental sailors aboard American ships. As a result, the navy discharged the Japanese messmen and replaced them with black sailors from other shipboard positions. In the minds of the naval commanders the action solved not one but two problems: It eliminated Japanese sailors and locked African-American sailors into the role of busboy and servant.

The lack of black recruits and a decrease in black reenlistments because of transfers to menial jobs steadily reduced the number of blacks in the navy. By the end of 1906 black sailors numbered only fifteen thousand in the entire U.S. Navy—less than 5 percent of the total force.

Over the next three years, the navy transferred additional black sailors to messmen and firemen positions until all but a few worked in the galleys or boiler rooms. In 1909 the navy removed all black petty officers from the fleet and assigned them to shore duties. African Americans in the navy shared the feelings of one of the transferred petty officers, Charles F. Parnell, who, after reporting to his land job, wrote, "Every one of us was transferred. We knew that the end of a colored man being anything in the navy except a flunky had arrived."

In addition to restricting African Americans to jobs belowdecks, the navy also made efforts to prevent black sailors from appearing in public. At the National Naval Review parade in New York City on October 12, 1912, not a single black was in the ranks of more than six thousand sailors who marched before the cheering crowds.

When African-American civilian leaders protested the absence of blacks in the formation, the navy, by direction of Congress, assigned Rear Adm. Charles J. Badger, commander of the Atlantic Fleet, to investigate charges of discrimination. Neither Badger nor the navy apparently found fault with the system of limiting black sailors to the enlisted ranks or restricting them to servant and laborer positions. An article in the May 26, 1912, edition of the *New York Times* reported that Badger and Secretary of the Navy Josephus Daniels "have completed an investigation of the charge of discrimination in the navy against colored enlisted men and find that the charge is unwarranted as there is no evidence of discrimination."

Despite their poor treatment and limited opportunities, African Americans actually fared better in the army and the navy than many senior politicians desired. Instead of being content with limiting black expansion in the army to four new bandmasters and allowing blacks to remain in the navy only as messmen and firemen, many American leaders worked to eliminate all African Americans from the U.S. armed forces. In December 1906, during the first congressional session after the Brownsville incident, Cong. John Nance Garner of Texas introduced a bill that called for eliminating all blacks currently in the military and barring black enlistments. Garner, whose district included Brownsville, introduced similar bills in each of the next three sessions of Congress. His zeal against blacks proved a political asset, for he won repeated elections to Congress and, as the running mate of Franklin Roosevelt, became vice president of the United States in 1932 and again in 1936.

Other congressmen joined Garner's anti-blacks-in-the-military movement. In each year between 1906 and 1916 someone introduced a bill that would eliminate the right of blacks to serve. When these bills failed, their sponsors proposed legislation to prohibit blacks from serving as officers or NCOs in the army and the navy. The War Department, citing the past "gallant service" of black soldiers and sailors in time of great need, expressed opposition to each of the bills. As a result, none ever reached the floor for a vote. However, the introduction of the bills themselves reinforced doubts in the minds of blacks that they could receive any degree of respect or equality in the military.

7

PRELUDE TO WORLD WAR: MEXICO AND HOUSTON

When war began in Europe in 1914, most Americans were relatively unconcerned about the possibility of U.S. involvement in the conflict. Supporting a policy of isolationism, American whites profited handsomely from the war by selling food, supplies, and raw materials to both sides, while black Americans concentrated on dealing with the extreme prejudice and discrimination that prevented equality and at times threatened their own safety.

Even though the war across the Atlantic was escalating, events closer to home were more pressing. Increased raids from Mexico by bandits and loosely organized groups of "revolutionaries" challenged the security of American residents in the Southwest and the sovereignty of the United States' borders. When white Americans could no longer tolerate the violations, they once more called upon the Regular Army black regiments. In rendering their services, black regiments participated in a pivotal period in U.S. military history—a time that brought both innovations to the battlefield and marked the end of some long-held traditions.

In 1912 the army ordered the black Ninth Cavalry to Douglas, Arizona, and in 1914 transferred the African-American Tenth Cavalry to Fort Huachuca, Arizona. Both regiments joined white units responsible for patrolling the border between Mexico and Arizona, New Mexico, and Texas. They spent their time on long, monotonous patrols along narrow trails in arid, harsh country. The occasional firefight did occur, but the border bandits usually fought only long enough to cover their escape. One of the few sustained battles occurred in March 1913 between a company of the Ninth Cavalry and about fifty Mexicans and Yaqui Indians near Douglas. While it produced only one dead bandit and no cavalrymen casualties, the fight represented a turning point in the Mexicans' willingness to stand and fight.

The root of the border conflict was Mexico's continued internal political upheaval; a series of violent changes in the government leadership precluded any faction from maintaining control of the country or its poor population. Some bandit groups touted allegiance to various political leaders, elevating themselves, in their eyes, from thieves to revolutionaries. When the United States recognized and supported the government in 1914 of Venustiano Carranza, who had taken control the previous year, the border incidents increased significantly.

One of the stronger revolutionary leaders was Pancho Villa, a former bandit who suffered several defeats by the Mexican federal troops in 1915. Blaming his failures on U.S. support of the current government, Villa believed that if he could bring the United States itself into the conflict, Mexicans would unite behind him both to fight the Americans and to expel Carranza. On January 10, 1916, Villa's "troops" stopped a train carrying personnel of the Cusi Mining Company at the cattle-loading station of Santa Ysabel, between El Paso and Chihuahua City. Shouting, "Viva Villa," the revolutionaries executed sixteen American miners.

The Ysabel Massacre failed to bring the United States to war in Mexico, so Villa took more direct action. During the early morning of March 9, Villa personally led five hundred men on a cross-border raid of Columbus, New Mexico, killing five Americans. The following day, President Woodrow Wilson ordered Brig. Gen. "Black Jack" Pershing to lead a fifteen-thousand-man punitive expedition into Mexico to kill or capture Villa. Wilson added difficult restrictions to the instructions, ordering Pershing to conduct his operations "with scrupulous regard to the sovereignty of Mexico."

Pershing quickly assembled combat regiments and support units along the border and on March 15 crossed into Mexico with his troops deployed in two columns, one originating from Columbus with the Eleventh and Thirteenth Cavalry and the other from Hachita, New

Mexico, with the Seventh and Tenth Cavalry. The two rapidly moving columns, supported by artillery and aerial observation supplied by the first American airplanes used in combat operations, converged at Colonia Dublan, about eighty miles south of the border. Pershing then organized three columns to continue south to catch Villa before he crossed the mountains into Sonora.

The black cavalrymen had greatly impressed Pershing during his previous service with them in Cuba. Confident of their abilities, Pershing assigned the Tenth Cavalry the responsibility for two of the three columns he dispatched from Colonia Dublan. For the next month, the Tenth Cavalry as well as the white regiments pursued Villa with little to show for their efforts except saddle sores and sunburns, for the countryside and its inhabitants protected the former bandit and his men. During the first twenty-eight days of the punitive expedition, the Tenth Cavalry rode more than 750 miles.

In one of the most successful engagements during the month-long pursuit, the Tenth Cavalry established an army "first." On April 1, F and H Troops encountered a band of 150 revolutionaries at a cluster of ranch buildings known as Agua Caliente. Maj. Charles Young, the third black graduate of West Point and the highest-ranking African American in the Regular Army, ordered his horsemen on line and directed his machine-gun detachment to set up on nearby high ground. Young and his cavalrymen then advanced under cover of the machine-gun fire, killing three of the enemy and scattering the rest. Agua Caliente marked the first use of machine guns in combat by U.S. soldiers.

Shortly after the battle at Agua Caliente, the Tenth received orders to reinforce the Thirteenth Cavalry at Santa Cruz de Villegas, where they engaged more than five hundred hostile Mexican federal troops, who resented the American incursion into their territory. Riding without halts, the Tenth arrived to support the Thirteenth after the white cavalrymen had killed forty of the Mexicans and lost two of their own. Although the casualty numbers were favoring the Thirteenth, the situation had been deteriorating before the arrival of the Buffalo Soldiers, which caused the Mexicans to withdraw. Some accounts credit the commander of the Thirteenth's troops, Maj. Frank Tompkins, with welcoming the Tenth by exclaiming, "By God, Young, I could kiss every black face out there."

On June 21 the Tenth Cavalry experienced its own battle with the federal troops. No one knows who fired the first shots when the black cavalrymen attempted to enter the village of Carrizal, but by the end of the fight twelve Americans lay dead, and eleven more were wounded. Estimates placed Mexican losses at seventy-five killed or wounded.

When the bodies of the dead Tenth cavalrymen arrived in El Paso, Americans honored them as heroes in one of the few public gestures of gratitude for black servicemen of the period. Large crowds also attended the burial of six of the men a few days later at Arlington National Cemetery.

After the Battle of Carrizal the Tenth continued its patrols in pursuit of Villa without success. Other regiments, including the Twenty-fourth Infantry, which joined the punitive expedition in May 1916, also failed to locate the outlaw.

Black and white cavalrymen secured food for themselves and fodder for their horses without benefit of supply lines on their long patrols. Col. William C. Brown, the commander of the Tenth Cavalry, described the difficulties of no supply lines as well as the problems of searching for a local bandit guerrilla fighter operating in his home territory in a May 1916 report to General Pershing. Brown wrote:

> I have drawn my personal check today for $1,100 gold to loan to officers to purchase supplies, forage, and rations. I had previously spent my last cent for this purpose. . . . Guides and messengers are hard to get as they fear retaliation by Villa. We have practically no horseshoes left. Men have lived on fresh beef, tortillas made from corn meal which we have ground ourselves. My personal opinion is that the various demoralized Villa bands will soon be scattered so that it will prove fruitless to follow them.

Finally, with the Mexican government protesting the U.S. invasion of its territory and an escalating war in Europe threatening to involve the United States, President Wilson ordered the expedition home. The last American units reentered the United States on February 5, 1917. Both the Tenth Cavalry and the Twenty-fourth Infantry returned to guarding the border.

Even though the U.S. abandoned its chase of Villa, the outlaw did not end his plan to antagonize the United States into commitment in

Mexico. He and his band made just enough minor attacks along the border during the following years to keep the patrols of the black regiments from being totally bored. In June 1919, Villa assembled a small army to attack Juarez; from there his troops fired a few shots into neighboring El Paso. Soldiers of the Twenty-fourth Regiment joined white troops in crossing the border and routing Villa's army. The Mexican bandit-revolutionary never again threatened U.S. territory, and in 1923 a rival Mexican faction assassinated him.

Although the punitive expedition had failed to catch Villa, the operation proved a benchmark in U.S. military history, for it was a time when the traditions of the horse cavalry's providing for its own needs without fixed supply lines overlapped the military experiment with mechanized vehicles and airplanes. After the punitive expedition into Mexico, the horse-mounted cavalry never again expected its units to "live off the land." From that time on, mechanized vehicles replaced horses as a mode of troop transportation and allowed for the implementation of supply lines from base camps to soldiers at the front. What Pershing and his subordinates learned about the use of mechanical vehicles, airplanes, and machine guns during the expedition would prove extremely useful in the impending involvement in the war in Europe.

Despite their earlier attitudes and policies, by 1917 Americans found themselves unable to remain bystanders in the European conflict that now engulfed the continent. The sinking of the *Lusitania* by a German submarine, the discovery of a letter from German foreign minister Arthur Zimmermann to his representative in Mexico City proposing an alliance between Germany and Mexico, and the disruption by German U-boats of international "shipping lanes," created threats to American security. On April 6, 1917 the United States declared war on Germany and its allies.

Black soldiers and sailors stood ready to serve overseas, and black communities were prepared to support the war effort. However, as in wars past, racism initially thwarted African Americans' opportunities to contribute.

Despite their outstanding service in Mexico and their long record of brave performance in previous wars, the four black regiments did not receive orders to Europe after the declaration of war because many senior military and political leaders still believed that African Americans lacked the intelligence, courage, and dedication to serve in sustained combat roles. These same leaders maintained that both the army and navy should limit blacks to menial positions as laborers and servants, and many continued to question whether blacks should serve at all. Even white supporters still believed that the blacks' "place" in uniform was certainly not as equals on the front lines or on the high seas.

The initial buildup to 750,000 men in the Regular Army and the National Guard at the beginning of World War I included only 20,000 black men, less than 3 percent. Only three blacks in the entire army held commissions.

Although the four black regiments contained the best-trained and most experienced soldiers in the U.S. Army, the War Department assigned them to continue their security duties along the Mexican border. Many of the regiments' soldiers believed that the military feared the consequences of having American black soldiers contribute to defeating a white enemy.

In July 1917 the Third Battalion, Twenty-fourth Infantry, received orders to report to Houston, Texas, to guard the construction of Camp Logan, where white troops were to train before transferring overseas. Black soldiers resented not being included in the units deployed to the European battlefront. Within a month pent-up frustrations led to the first racial riot in American history in which more whites died than blacks.

The Third Battalion arrived in Houston in late July angry that the army had excluded them from the war in Europe but happy to be assigned near a large city whose black population constituted nearly a quarter of the total residents. Meanwhile, Houston's civic leaders also had mixed emotions. While they welcomed the income potential from Camp Logan, they were most displeased to suddenly find a battalion of armed black men in their midst.

Normally and ideally, both white and black citizens would have been able to rely on the local police force to adjudicate differences. But in the case of Houston in 1917, the police force itself was a third party to inevitable problems. With a total number of 159 members, only two of whom were black, the police department generated trust from none of the city's citizens, regardless of color. In fact, an article in the February 22, 1917, edition of the *Houston Chronicle* professed, "Houston seems like a city without a police force."

From the outset of their arrival, black soldiers were subject to police harassment, beatings, and arrests for not honoring the local Jim Crow laws, which included segregated seating on city streetcars. Even the black military police fell under the arbitrary nature of white law-enforcement officials, who demanded that the M.P.s leave their pistols on the post and carry only billy clubs when on civilian streets, requirements that the soldiers resented as a lack of respect.

The soldiers of the Twenty-fourth found the situation no better at the Camp Logan construction site, where white workmen openly called them "niggers" and "coons." Separate water cans and toilet facilities further divided the black guards from the camp's white workers. As bad as it was in Houston, however, the conditions were no worse than those that the Twenty-fourth had endured at other Texas and western outposts.

Some observers later theorized that the African-American soldiers of the Twenty-fourth Regiment had finally experienced too much prejudice and disrespect in Houston, while others thought that the night of violence that followed resulted from the culmination of nearly three centuries of abuse. While these factors certainly played a role, several others beyond the usual racism and discrimination significantly contributed to what became known as the Houston Mutiny.

The Twenty-fourth itself was a different unit than it had been before its transfer to Houston. While the Third Battalion had a full complement of enlisted men, fifty of them were new recruits who joined the unit shortly before its move. More importantly, most of the senior NCOs had been reassigned to the Colored Officer's Training Camp at Des Moines, Iowa, to earn commissions in the newly authorized black divisions preparing for transfer to the European battlefield. The battalion's sergeant major, the highest-ranking NCO, and all the company first sergeants in the unit except one had been among the twenty-five senior NCOs transferred to Des Moines.

Because the discipline and morale of the Twenty-fourth, like any other unit, depended on the effectiveness of the noncommissioned and commissioned officers, the reassignment of these experienced leaders left less experienced men in key roles. Within battalions and regiments the NCOs controlled the discipline and ran the day-to-day activities of the units. Although the commissioned officers did command, most of their efforts focused on relations with higher commands and on operations planning. In the black regiments it was the duty of the black NCOs to maintain control of the enlisted men; white officers, in addition to their regular duties, protected the welfare of the men in their relationships with the local communities.

Further altering the composition of the Twenty-fourth was the change of commanders which occurred several weeks after the Twenty-fourth's arrival in Houston, when the Third Battalion's extremely professional commander, Lt. Col. William Newman, was transferred to a new assignment and the newly promoted and far less capable Maj. Kneeland S. Snow assumed command on August 20. Snow preferred socializing in Houston to spending time with his troops at Camp Logan. As a result, Capt. Haig Shekerjian, who assumed the duties of battalion adjutant on August 6, handled most of the routine command matters. Compounding the chain-of-command difficulties, too, was that inexperienced lieutenants instead of veteran captains were filling half of the company command positions.

In Houston, the black NCOs and the white officers failed to properly maintain their regiment's morale and discipline. Some of this failure came from the personalities of the participants, but the major breakdown resulted from the chain-of-command changes and nature of the times. Even the weak NCO and officer structure and the long years of discrimination might not have led to a full-scale riot between the black soldiers and the local white citizens had not a particularly abusive incident involving the Houston Police Department set it off.

On the morning of August 23, 1917, Houston policemen arrested Sara Travers, a black housewife and mother of five children, for complaining about two patrolmen shooting into her neighborhood while attempting to apprehend two young crapshooters. One of the policemen, Lee Sparks, angrily told Travers, "You all God damn nigger bitches, since these God damn sons of bitches nigger soldiers came here, you are trying to take the town."

Sparks and his partner physically grabbed Travers and dragged her to a nearby call box, from which they requested a patrol wagon to transport her to jail. While the policemen and Travers stood waiting, a group of black civilians gathered to complain about the mistreatment of Travers. Pvt. Alonzo Edwards, of the Third Battalion's L Company, joined the group and offered to pay whatever fine was involved if the police officers would release the woman. Even though Edwards made no physical move against the policemen, Sparks drew his revolver and struck the black soldier several times on the head, later explaining, "I wasn't going to wrestle with a big nigger like that. I hit him until he got his heart right."

Sparks then took Edwards as well as Travers to the city jail. Early that afternoon, Cpl. Charles W. Baltimore, a senior member of the battalion's provost guard, went to the jail to check on Edwards's welfare. Although evidence indicates that Baltimore inquired politely, Sparks again drew his gun and stuck the black corporal over the head. The unarmed Baltimore ran from the jail as Sparks fired three shots at him and then chased the fleeing soldier. Sparks, a better runner than marksman, caught Baltimore, struck him several more times over the head with his gun, and placed him in the city jail along with Edwards.

By midafternoon rumors began that the local police had murdered Baltimore. Major Snow sent Shekerjian to Houston to investigate. The adjutant secured Baltimore's release and a promise that Edwards would also be freed as soon as clean clothes arrived to replace his blood-soaked uniform. Shekerjian also convinced the police chief to investigate Sparks's actions.

By the time Shekerjian and Baltimore returned to camp, the rumors that police had killed the corporal were rampant. Stories spread that the white citizens of Houston planned to raid the soldiers' camp that evening. Discovering that Baltimore was alive but severely beaten did not diminish their outrage. Snow, however, believed that the crisis had passed and took no further action except to restrict his battalion to camp.

Anger continued to build, and by early evening soldiers began assembling and shouting that the time had come to take their revenge. Although several of their white officers attempted to stop them, about a hundred soldiers marched out of Camp Logan into the streets of Houston at about 8:40 P.M., appearing more mob-like than military in their formation. What organization that did exist seemed spontaneous rather than pre-planned. Several NCOs joined the formation, including a battered and humiliated Corporal Baltimore, who sought revenge on the Houston Police Department. By design or because of his rank, Sgt. Vida Henry, the acting first sergeant of I Company, took command of the group.

Over the next two hours the soldiers proceeded to deliberately shoot and kill policemen as well as any civilian bystanders who got in their way. Destroying no private property, they remained focused on their targets—members of the Houston police force. But their bullets did not always find their intended marks. In addition to several civilians, the mob mistakenly identified white army Capt. Joseph Mattes and one of his men of the Illinois National Guard as policemen and fired more than fifty shots into their vehicle.

After two hours of combing the Houston streets for their prey, the soldiers broke into small groups and either returned to Camp Logan or sought refuge with local families. In their wake they left fifteen dead, including four policemen, two soldiers mistaken for policemen, and two civilians who had been assisting the police. The mob wounded twelve more white Houston citizens, including a policeman, who would die of his injuries. Officer Sparks was not among the dead or wounded.

Four black soldiers died during the mutiny, including Sergeant Henry, who committed suicide with his army-issue rifle after the shooting ceased. Neither Henry's last words, "I ain't going in, I ain't going to camp no more," nor any other source ever explained why he led the group into Houston. Some investigations later credited him with planning the entire mutiny, while others claimed that Henry participated only out of a sense of responsibility to his men and then killed himself rather than live with the consequences.

The morning after the riot, the Third Battalion officers disarmed all of their soldiers and began comparing roll-call figures to identify the men who had missed formations held during the night. Various reports did not initially match, but finally the officers agreed that 156 men had missed roll call on the night of August 23.

On the morning of August 25 the Third Battalion, Twenty-fourth Infantry, boarded trains for transfer to remote Camp Furlong, near Columbus, New Mexico. Upon arrival at Camp Furlong, the 156 men who had missed roll call were placed in a stockade pending further investigation.

Unlike its actions following the Brownsville incident, the army did not resort to mass punishment. Extensive investigations both by the army and the city of Houston attempted to discover exactly what had happened and who the individual guilty parties were. The investigations cited the arrest and pistol whipping of Private Edwards and Corporal Baltimore by Officer Sparks—and the subsequent rumors of Baltimore's death—as the flash point for the mutiny, but not a single witness in Houston could identify any of the soldiers who shot the victims. Finally, after granting immunity to several of the participants and comparing roll-call rosters and observations by the battalion's officers in camp, the army identified individual defendants and assembled its charges against each.

The first court-martial convened at Fort Sam Houston in San Antonio, Texas, on November 1 to try sixty-three of the soldiers charged with mutiny and premeditated murder. The trial, the largest military court-martial in U.S. history, involved 196 witnesses and resulted in twenty-one hundred pages of typewritten testimony. On November 12 the board pronounced fifty-four of the defendants guilty of all charges. Of those, thirteen received death sentences, and forty-one were sentenced to life in prison. Four soldiers received lesser sentences, and five went free.

Before dawn on December 11, 1917, army executioners hanged Corporal Baltimore and twelve other black soldiers at Camp Travis, just outside Fort Sam Houston, without notice to the public or to the press. Not until two hours after the executions did the army announce the fact.

Neither the War Department nor the president had reviewed the trial findings. When black civilian leaders protested the lack of high-level review, the army responded that it had acted properly in enforcing the sentences, "since military law specified that the area commander had final authority in time of war."

When black leaders and newspapers protested to President Wilson and the War Department that Texas and the trials were far from the European war, they were able to secure a promise of high-level review of any future death sentences.

After the first trial, the white press in the South and the North mostly agreed that justice had been served. The editor of the *Atlanta Constitution* wrote in his December 13, 1917, edition, "Justice has been done at Houston, not summary, hair trigger, hasty justice, but deliberate, mature, carefully weighed and unbiased justice."

Opinion among the black press ranged from regret to anger. The December 15, 1917, edition of the *Chicago Broad Ax* expressed regret that the Twenty-fourth Infantry Regiment had "blown their past splendid record to the winds." C. N. Love, the editor of the *Houston Observer*, expressed regrets about the incident but noted: "The chances of it occurring would have been remote if a disciple of 'democracy' [Officer Sparks] had not overstepped his bounds in dealing with a black soldier."

W. E. B. Du Bois expressed perhaps the mostly widely felt belief in the black community about the Houston Mutiny in the January 1918 edition of the *Crisis*:

> They have gone to their death, 13 young, strong men; soldiers who have fought for a country which never was wholly theirs; men born to suffer ridicule, injustice, and, at last, death itself. They broke the law. Against them punishment, if it was legal, we cannot protest . . . the shameful treatment which these men, and which we, their brothers, receive all our lives, and which our fathers received, and our children await; and above all we raise our clenched hands against the hundreds of thousands of white murderers, rapists, and scoundrels.

Two subsequent trials tested the review system as sixteen additional members of the Third Battalion received death sentences and twelve received life terms. After strong lobbying from African-American leaders, President Wilson, in an effort to appease both the black and the white communities, changed the sentence of ten of the soldiers to life in prison and authorized the execution of the remaining six on September 16, 1918.

Most of the press interest in the Houston Mutiny occurred only during and after the first trial. By the time of the second and third trials and executions, much more of the press's efforts focused on the American military involvement in France.

Over the years continued lobbying by black organizations, including the National Association for the Advancement of Colored People (NAACP), secured reductions in many of the sentences. However, it was not until twenty years later, in 1938, that the final Houston mutineer was released from the federal prison at Fort Leavenworth, Kansas.

The Third Battalion, reinforced by recruits, served at several isolated forts in New Mexico before the army disbanded it altogether in 1921. The remainder of the Twenty-fourth Infantry Regiment served similar postings, remaining beyond consideration for favorable assignments for the next two decades before the magnitude of World War II required an entirely new look at the use of black units.

Unlike Brownsville, where the black soldiers received mass punishment rather than fair trials, the army adjudicated the Houston Mutiny through proper procedures. Certain soldiers of the Third Battalion mutinied against their officers and murdered local civilians. While a long history of discrimination and mistreatment may have provided the background against which their actions took place, the fact remained that there was no justification for the soldiers to take the law into their own hands. Although the darkest days of African-American service in the uniform of the United States had passed, the impact of those two violent, deadly hours on the streets of Houston would haunt black soldiers and the U.S. Army for decades to come.

8

THE GREAT WAR: WORLD WAR I

The entry of the United States into World War I on April 6, 1917, stimulated the same debates about African-American participation as had previous conflicts. Within the black community, some factions argued the advantages of recognition and equality that they could achieve through military service, while others cited failures of the United States to significantly reward blacks for their earlier wartime service. In the meantime, few whites expressed enthusiasm for additional black enlistees. However, true to centuries of history, once the war escalated to the point of requiring great numbers of soldiers, African Americans once again found themselves welcomed into the ranks, albeit into segregated units commanded mostly by white officers.

The debate among black leaders over military support began during the prewar buildup and continued after the entry of the United States into the conflict. Most of this discussion took place through public forums and did not appear in print until later. Columbia University law student Randolph Owen summarized the position of those who opposed black enlistments in an article that appeared in the January 1918 edition of the *Messenger*. Owen wrote: "Since when has the subject race come out of a war with it rights and privileges accorded for such participation? . . . Did not the Negro fight in the Revolutionary War . . . and come out to be miserable chattel in this country for one hundred years after? . . . Did not the Negro take part in the Spanish-

American War? . . . And have not prejudice and race hate grown in this country since 1898?"

Owen presented valid questions, but most black leaders knew that resistance to supporting the war would do nothing to advance the status of their race. Some openly expressed their enthusiasm about participation in fighting and its rewards. At Howard University a student and faculty organization issued a proclamation declaring, "If we fail, our enemies will dub us cowards for all time; and we can never win our rightful place. But if we succeed—then eternal success."

The most eloquent and influential spokesman supporting African-American service in World War I, indeed for all racial issues of his generation, W. E. B. Du Bois, argued that educated blacks had to take the lead. This "talented tenth," according to Du Bois, should lead the black community in support of the war. In an editorial in the June 1918 issue of the *Crisis*, Du Bois concluded that these actions would result in "the right to vote and the right to work and the right to live without insult."

In the next month's edition of the *Crisis*, Du Bois called on African Americans to put aside their "special grievances" for the rest of the war and to "close our ranks shoulder to shoulder with our white fellow citizens and the allied nations that are fighting for democracy. . . . We make no ordinary sacrifice, but we make it gladly and willingly with our eyes lifted to the hills."

Although African Americans were in the end willing to put aside their "special grievances" out of a sense of patriotic duty and a hope for gaining further equality, white America did not rush to accept black volunteers into uniform. During the prewar buildup of 1916, the army recruited 650,000 volunteer soldiers but limited the number of black enlistees to only 4,000, all of whom were assigned to the four Regular Army regiments serving in the Southwest. When those regiments reached full strength, the army ceased accepting black volunteers because all the new units being formed were "whites only."

Not until the United States actually entered World War I and calculated the need for 4 million volunteers and draftees to man its forces did white America look to the black community for potential recruits. The military once again lifted its restrictions on African-American enlistments because of the dire need for personnel.

Provisions of the Selective Service Act required that every able-bodied male from twenty-one to thirty-one years of age, black or white, register for the draft. Ironically, it was the prejudiced whites serving in the Selective Service System who unwittingly advanced black equality. Local draft boards administered the system and reviewed the registration of each young man in their town or district for induction. Draft boards across the country recognized the chance not only to defer the white sons of the wealthy and influential but also to "rid" their communities of young African Americans.

Many draft boards required black men to tear off a corner of their registration forms so that they were more readily identifiable. As a result, the boards drafted a disproportionate number of African Americans. During the war, a total of 2,291,000 black men registered for the draft, about 9 percent of all registrants. The Selective Service System drafted 367,710, or 13 percent, of the black men registered.

With more than a third of a million black draftees, the War Department faced the dilemma of what to do with the new African-American soldiers. Original plans to form sixteen black infantry combat regiments quickly changed with the news of the Houston Mutiny of August 1917. White politicians and their supporters expressed concerns about further arming and training large numbers of African Americans. As a result, the War Department reduced the number of proposed black combat regiments from sixteen to four and assigned the remainder of the black draftees to Services of Supplies (SOS) units. Instead of joining the prestigious ranks of combat soldiers, most black draftees would serve as cooks, drivers, and laborers.

Although blacks' entry into the U.S. military at the beginning of World War I proved easier than in previous wars, once again African Americans had to fight for the right to fight. Black leaders recognized that assignment to menial service jobs would do little to prove the worth of the black soldier and would produce few postwar benefits of value. African-American groups across the United States, including the NAACP, lobbied for additional black combat troops as well as the opportunity for blacks to earn officer commissions.

Criticism of the government's policies on reducing the number of black combat units and providing commissions for black officers reached such a level that President Woodrow Wilson and Secretary of War Newton D. Baker had to take action. Less than a month after the Houston Mutiny and the announcement of the reduction in planned black combat regiments, Secretary Baker appointed Emmett J. Scott, Booker T. Washington's personal assistant for eighteen years before the leader's death in 1915, as his special adviser on black affairs. Included in Scott's duties was the task "to be responsible for all cases of real or alleged discrimination against Negroes in the army."

Some black leaders believed that Scott's appointment was a token gesture, suspecting the special adviser of being too much of a lackey of the white man's system. Nevertheless, Scott proved to be a tireless worker for the rights of black soldiers, handling thousands of complaints about racism in the ranks and by draft boards. Scott also established programs that led to officer training in more than twenty African-American colleges and universities and exerted influence that increased commissioning opportunities for black enlisted men.

On June 15, 1917, the army opened a training camp for black officer candidates at Fort Des Moines, Iowa. The first class of 639 graduated the following October, and by the end of the war, twelve hundred African Americans had earned commissions at the school. More than 250 of the officer candidates came from the NCO ranks of the four Regular Army regiments. The remaining volunteers reported from National Guard units and from the civilian talented tenth that Du Bois had encouraged to step forward.

Although the total of more than twelve hundred commissions for black soldiers far exceeded the number from previous wars, it was hardly in proportion with the overall numbers of African Americans in uniform. While black enlisted men made up 13 percent of the total active-duty force, black officers represented only seven-tenths of 1 percent of the officer corps. Black officers faced promotion restrictions in addition to numerical restrictions. Most remained lieutenants and captains; only a few advanced to the field-grade ranks of major and above.

One of the primary reasons for not promoting blacks beyond the grade of captain was to limit the chances of white soldiers being subordinated to them. Within the prewar army, only in the regular black regiments did a few black officers outrank whites with commissions. In some cases, white officers requested and were granted transfers when placed subordinate to blacks.

Many in Washington and within the army found the idea of black senior officers so repugnant that they went to extraordinary lengths to block African Americans from achieving large commands. As the army mobilized for the war in Europe, black leaders and newspapers lobbied for the promotion of Lt. Col. Charles Young to full colonel and that he be given command of his own regiment, two actions that would place him in line to become a general officer. None of these options found favor with white commanders. Instead of promoting Young, the War Department announced that he had high blood pressure and medically retired him from active duty. To appease the many critics of the move, the army promoted Young to colonel on the retired rolls, meaning additional pay but no position in the active military.

Young, feeling hurt as well as healthy, did not readily accept retirement. He rode horseback from his home in Chillicothe, Ohio, to Washington, D.C., to prove his good health and to apply for reinstatement to active duty. The War Department denied his request. Only a few days before the war ended, the army finally reconsidered and recalled Young to active duty—to serve as an adviser, with no command responsibility, to the Ohio National Guard.

Despite protest from black leaders, the army continued to assign the vast majority of African-American volunteers and draftees, along with officer graduates of Des Moines, to labor and stevedore battalions instead of combat regiments. Black soldiers landed in France with the initial elements of Gen. John J. Pershing's American Expeditionary Force (AEF), but instead of carrying rifles or manning artillery pieces, these early arrivals unloaded ships, stacked supplies, drove trucks, and built tent cities.

Ultimately 160,000 of the 200,000 black soldiers deployed to France served in SOS units rather than in combat assignments. Not only did they have the dirtiest, meanest manual-labor jobs, including grave digging, but also the black labor battalions were last in priority for housing, rations, uniforms, and other supplies.

The 20 percent of black soldiers, or forty thousand, who did see combat in World War I served with the Ninety-second and Ninety-third Infantry Divisions. Four newly authorized infantry regiments—the 365th, 366th, 367th, and the 368th—constituted the bulk of the Ninety-second. Two machine-gun battalions and additional artillery, engineer, signal, and other support units rounded out its organization.

Most of the enlisted manpower of the Ninety-second Division were draftees, and the majority of the junior black officers reported directly

Members of the 317th Engineers on construction detail near Varennes, France, on October 25, 1918. (U.S. ARMY MILITARY HISTORY INSTITUTE)

from the Fort Des Moines training center. From its very beginnings, it appeared that the War Department set up the Ninety-second Division to fail. Because white communities feared large numbers of black combat troops, the Ninety-second never assembled as a division to conduct training before deploying for Europe. The infantry regiments and support units trained at seven widely separated camps and had no interaction until actually landing in France.

In addition to never training his division as a whole, the Ninety-second's commander, Maj. Gen. Charles C. Ballou, failed to take any positive stand to deter the discrimination that his soldiers faced in the communities adjacent to their camps. Selected for the command because of his previous experience in the Twenty-fourth Infantry Regiment and on the staff of the Fort Des Moines training center, Ballou rarely stood up for his men. He would place the burden of proof for their defense on his black soldiers when there was a dispute with white military personnel or civilians.

At Camp Funston, Kansas, the training site for one of the Ninety-second's regiments, a black soldier in early 1918 protested being denied entry into a theater in nearby Manhattan. Although Kansas law pro-

Infantrymen of the Ninety-second Division's 366th Regiment test gas masks at Ainvélle, France, on August 8, 1918. (U.S. ARMY MILITARY HISTORY INSTITUTE)

hibited such discrimination, Ballou sided with the theater owner and ordered his men not to go where they "were not wanted." Ballou issued Bulletin No. 35 to all members of the division outlining his ideas on relations between blacks and whites: "The Division Commander repeats that the success of the Division, with all that success implies, is dependent upon the good will of the public. That public is nine-tenths white. White men made the Division, and they can break it just as easily if it becomes a trouble maker."

While the Ninety-second Division organized its draftees into combat units and conducted what training it could under the circumstances, the War Department faced the challenge of assigning the black National Guard units that had joined the war's mobilization. Initial plans were to assign the black units to divisions formed by white National Guard soldiers; white division commanders and political leaders from the activated units' states expressed such displeasure at the inclusion of African Americans that the War Department relented.

With politics and racism once again winning out over logic and military requirements, the War Department reluctantly authorized the formation of a second black division, the Ninety-third. Unlike the 92nd, however, the Ninety-third contained no service or support units, making it incapable of independent operations.

The entire Ninety-third Division consisted of only four infantry regiments. Three of these came from black National Guard units from Connecticut, Illinois, Maryland, Massachusetts, New York, Ohio, Tennessee, and the District of Colombia. A fourth regiment, composed of draftees, rounded out the troop list of the Ninety-third.

Like the Ninety-second, the regiments of the Ninety-third also trained at separate, distant installations before and after assignment to the new division and received no better welcome from nearby communities than did the Ninety-second. When the Fifteenth New York, soon to be redesignated the 369th Infantry Regiment, reported to Camp Wadsworth, South Carolina, in October 1917, local shopkeepers in nearby Spartanburg refused to serve the black soldiers, and the townspeople cursed and forcibly removed a black officer—a Harvard University–educated lawyer—from a "whites only" streetcar. In another incident, a white hotel keeper knocked down and kicked a senior black NCO for not removing his hat in his presence.

Once again, the War Department failed to make a stand in defense of its black soldiers. Instead of placing any demands on the citizens of Spartanburg, the War Department ordered the New York regiment to board transports for Europe only two weeks after its arrival in South Carolina.

Because of the racism that prevented their training together as units, both the Ninety-second and Ninety-third Divisions arrived in France poorly prepared and under the leadership of inexperienced officers and NCOs. However, once in France, African Americans from the combat divisions and the SOS units discovered that the French treated them far better than had their own countrymen. White commanders expressed concern over the equality and friendliness the French extended to black soldiers and made efforts to stop such treatment.

On August 7, 1918, the leader of the French Military Mission to AEF headquarters sent a message, marked "Secret Information Concerning Black American Troops," to French military and civilian leaders. The French liaison officer, upon the advice of his American comrades, wrote:

> The American attitude upon the Negro question may seem a matter for discussion to many French minds. But we French are not in our province if we undertake to discuss what some call "prejudice." American opinion is unanimous on the "color question" and does not admit of any discussion. . . . It is of the utmost importance that every effort be made to avoid profoundly estranging American opinion. Although a citizen of the United States, the black man is regarded by the white American as an inferior being with whom relations of business or service only are possible. The black is constantly being censured for his want of intelligence and discretion, his lack of civic and professional conscience and for his tendency toward undue familiarity.

The message continued with an admission that military authorities could not intervene directly with actions by the civilian population but recommended that the locals not "spoil" the black soldiers and that they make particular efforts to prevent white French women from socializing with African Americans. In his conclusions the French liaison recommended conduct by the French military:

> We must prevent the rise of any pronounced degree of intimacy between French and black officers. We may be courteous and amiable with these last, but we cannot deal with them on the same plane as with the white American officers without deeply wounding the latter. We must not eat with them, must not shake hands or seek to talk with them outside of the requirements of military service. We must not commend too highly the black American troops, particularly in the presence of (white) Americans. It is all right to recognize their good qualities and their services, but only in moderate terms, strictly in keeping with the truth.

By the time the Ninety-third Division arrived in France in December 1917, its National Guard units and draftees had been reorganized into the 369th, 370th, 371st, and 372nd regiments. Because the division contained no service or support units, American commanders were at a loss as to how to employ the black regiments until the French, in desperate need of fresh troops, requested assignment of the regiments of the Ninety-third Division to their army. Pershing and his staff were happy to rid themselves of the problem and made the transfer with so

369th Infantry Regiment of the Ninety-third Infantry Division on review with French equipment at Maffrecourt, France, on May 5, 1918. (U.S. ARMY MILITARY HISTORY INSTITUTE)

much haste that Col. William Hayward, commander of the 369th Infantry Regiment, wrote a friend, "Our great American general simply put the black orphan in a basket, set it on the doorstep of the French, pulled the bell, and went away."

Despite warnings from U.S. headquarters about possible problems in dealing with black troops, the French, who had employed black soldiers from their colonies in the Sudan and Senegal with great success, welcomed the four regiments. They issued the black soldiers French rifles, French helmets, and French rations—only their uniforms remained American made—and treated them as equals, which was heretofore unknown by African Americans serving in U.S. commands.

The four regiments joined the French frontline forces in the early summer of 1918 without ever uniting under their own command and remained with them until the end of the war, participating in the Meuse-Argonne campaign and the Oise-Aisne offensive. Black soldiers learned their lessons on the battlefield and proved themselves equal to their Allies and a fearsome opponent of their enemies. Casualties in the Ninety-third's four regiments totaled 584 killed and more than twenty-four hundred wounded, for a casualty rate of 35 percent of its assigned strength. About 550 officers and men of the Ninety-third received combat decorations from either or both the French and American armies.

All four of the regiments performed well in combat, with the 369th leading the way by earning the French croix de guerre for their gallantry at Maison-en-Champagne. The 369th, formed from the Fifteenth New York, became known as the "Men of Bronze" to the French and as "Hell Fighters" to their German opponents. In addition to their unit recognition, 171 officers and men of the regiment received individual awards of the croix de guerre or the Legion of Merit.

Regimental commander Colonel Hayward made every effort to promote morale within the 369th and to ensure the welfare of his men. Hayward often took a place in the mess-hall serving line to see that his soldiers received good food and to listen to their remarks and complaints. Before leaving New York, Hayward had assisted Lt. James Europe in recruiting some of the finest black musicians in the Northeast and secured funds from regional business leaders to buy instruments to form the regimental band. In addition to increasing morale in the regiment, Europe and his musicians became known as the band that introduced jazz to France.

The Men of Bronze, or Black Rattlers, as they called themselves, spent 191 days on the front line, longer than any other American regiment, black or white. During that time, the 369th never had a single soldier taken prisoner by the Germans, and the regiments never surrendered a single foot of ground to the enemy.

Among the heroes of the 369th were Pvts. Henry Johnson and Needham Roberts. Shortly after the regiment took up positions in the front-line trenches, Johnson and Roberts manned a night observation post forward of their company's position to provide early warning of attack. At 2:00 A.M. on the morning of May 14, 1918, a German patrol of twenty-four men, seeking prisoners from whom to gain intelligence, struck the observation post. Although Johnson and Roberts suffered

369th Infantry Regiment, Ninety-third Infantry Division using French weapons and equipment in the trenches near Maffrecourt, France, in 1918. (U.S. ARMY MILITARY HISTORY INSTITUTE)

severe wounds, they refused to surrender or retreat from their position and fought off the attack, using their rifles, hand grenades, and finally their bayonets. In recognition of their bravery the two men received the first two croix de guerre awarded to the Ninety-third Division by the French government.

Black soldiers of the Ninety-third Division's 370th Infantry Regiment also made their mark while serving as a part of the French army. The 370th had the advantage of six weeks of training in the United States before sailing for Europe and arrived better prepared than the division's other regiments. Formed from the veteran Eighth Illinois National Guard Regiment and composed mostly of men from the Chicago area, the 370th was the only regiment in the AEF led almost entirely by black officers, including its commander, Col. Franklin A. Denison.

After its arrival in France, the 370th joined the French Tenth Division and occupied a portion of the line at St. Mihiel. They then moved to the Argonne Forest and during the final months of the war participated in the advance in the Soissons sector that captured the German fortifications at Mont des Signes. The 370th performed adequately, but not as well as they might had Denison remained in command. Before the final Allied offensives, the U.S. Army replaced Denison, because of ill health, with Col. T. A. Roberts, the first white officer to command the Illinois National Guard soldiers.

The 371st, black draftees led by white officers, and the 372nd, black national guardsmen from Connecticut, Maryland, Massachusetts, Ohio, and Washington, D.C., led by white senior and black junior officers, arrived in France in the early summer of 1918 and joined the French 157th "Red Hand" Division. Both regiments suffered casualties similar to the Ninety-third Division's other two regiments in the war's final battles.

Unlike the Ninety-third Division, which attached its regiments to widely separated French commands and never fought as a unit, the Ninety-second Division retained its autonomy. Although poorly trained, the Ninety-second took over the St. Die sector from the U.S. Sixth Infantry Division on August 25, 1918. They immediately faced a German counteroffensive to capture the village of Frapelle, which included a propaganda campaign of leaflets delivered by artillery shells questioning the black soldiers about fighting for a country that denied them equality.

The commander of the Ninety-second, Maj. Gen. Charles Ballou, was not happy with his division's fighting abilities at St. Die, nor was he pleased with their aggressiveness or response to orders in the Meuse-Argonne offensive of September 26 to October 5. During that battle, the 368th Regiment became so confused and disorganized that Ballou had to withdraw them from the line. In a tradition as old as armies themselves, Ballou responded not by looking at the lack of training and experience of his division's regiments but by blaming his subordinates. He ignored the fact that when sent forward, the 368th had neither detailed orders nor essential supplies or equipment, including maps, hand grenades, signal flares, and wire cutters. Instead, Ballou placed particular blame on his newly commissioned black officers and court-martialed thirty lieutenants and captains for cowardice. An all-white court-martial convicted four of the officers and sentenced them to death, later reduced to long prison sentences by a review board.

The others had not yet come to trial when AEF headquarters transferred the Ninety-second to the control of the U.S. Second Army, commanded by Lt. Gen. Robert L. Bullard. As a former officer in the Third Alabama Volunteer Regiment during the buildup for the Spanish-American War, Bullard appreciated the capabilities of black soldiers and knew that any military organization required good leadership and training to be proficient.

Bullard made some progress in raising the morale and training level of the Ninety-second before the Allies resumed the conflict's final offensive. Two of the regiments of the Ninety-second, the 367th and the 368th, performed well in the war's last days, managing to push the Germans back, defeat several counterattacks, and provide support to a French unit on their flank. For its bravery in the drive against Metz on November 10–11, which halted only with the announcement of the armistice, the First Battalion of the 367th received the French croix de guerre.

Unfortunately, neither the 365th nor 366th Regiments ever gained a level of expertise equal to that of the other two regiments. Instead of fighting, both regiments ended up on road-building details to expedite supplies going to the front lines.

Despite the triumphs of the 367th and 368th, Bullard was disappointed with the performance of the rest of the division and concerned by how it reflected on his leadership. He removed Ballou from command and requested that the Ninety-second be transferred from the Second Army. Then, when the armistice ended the war before the transfer could take place, Bullard recommended the immediate transfer of the division back to the United States. In his written rationale for the urgent action of removing the Ninety-second, Bullard stated that it would be impossible to prevent black soldiers from assaulting French women if they remained in the country, even though only one such incident had been reported during the entire time the Ninety-second had been in France. The French commander, Marshal Ferdinand Foch, who respected the professional support provided his army by the regiments of the Ninety-third Division, saw through Bullard's thinly veiled racism and informed General Pershing that no combat forces should leave the country until a formal peace treaty had been signed.

Overall the French were pleased with the performance of the black regiments attached to their divisions. They also recognized the massive amount of labor provided by the large number of black SOS units that supported Allied combat troops.

General Pershing, as commander of the AEF, praised the actions of all black soldiers during the war but shared the concerns of his staff and subordinate commanders about the general poor performance by the Ninety-second Division. Neither Pershing nor his staff admitted that the primary difficulties in the Ninety-second's performance lay in the division's lack of training and with its senior white officers, who neither respected nor cared about their African-American subordinates. Moreover, in evaluating the effectiveness of black units, American officers failed to credit the Ninety-third Division's fighting ability and to acknowledge that it succeeded because its regiments came mostly from National Guard units that had served and trained together for years before their activation. Instead, they claimed that the regiments of the Ninety-third fought well only because they were interspersed with, and supported by, white French units.

The senior command also overlooked the individual proficiency and bravery of the black soldiers of the Ninety-second and Ninety-third Infantry Divisions. Not a single black soldier received the Medal of Honor for service in World War I until seventy-two years after the conflict. In response to research by historians, who noted the bias against awards for black soldiers in the war, President George Bush explained that the action was "long overdue" when he awarded a Medal of Honor to family members of Cpl. Freddie Stowers at a White House ceremony on April 24, 1991.

The action for which Stowers posthumously received the nation's highest military honor occurred on September 28, 1918, in the Champagne Marne sector of France. A squad leader in Company C, 371st Infantry Regiment, Ninety-third Infantry Division, Stowers participated in an attack on Hill 188 in which his company was the lead element. Within minutes after the Americans launched the assault, the enemy ceased firing and began climbing onto their trench parapets, holding up their arms as if to surrender. The Americans responded by ceasing their fire and advancing into the open to surround the would-be prisoners. Just as Stowers's company got to within 100 meters of the enemy, the Germans jumped back into their trenches and showered the Americans with interlocking bands of machine-gun and mortar fire. Company C took 50 percent casualties within seconds.

With his officers and NCOs wounded or dead, Corporal Stowers took charge of the survivors. Facing a determined enemy and a devastating wall of fire, Stowers rallied the men. He then led them forward toward the machine-gun position which was killing his comrades. With little regard for his own safety, Stowers directed a fierce fight against the position until its German defenders lay dead. Turning his men toward a second trench line, the African-American corporal once more prepared to lead the attack. As he crawled forward, urging his men to follow, machine-gun bullets ripped through his body. Still Stowers tried to continue the advance despite his mortal wounds. When he could go no farther, Stowers shouted encouragement to his men, propelling them forward with his last breath.

Inspired by the heroism they had just witnessed, the survivors pressed the attack and overran the remaining entrenched Germans, securing Hill 188 and inflicting heavy casualties on the enemy. The young corporal's long-overdue Medal of Honor citation concluded: "Stowers's conspicuous gallantry, extraordinary heroism, and supreme devotion to his men were well above and beyond the call of duty, follow the finest traditions of military service, and reflect the utmost credit on him and the United States Army."

The army also did not recognize the individual technical skills possessed by African Americans. Although World War I provided blacks their first opportunity to serve in all enlisted positions in the infantry divisions, including the previously excluded artillery branch, the army once again did not allow them to participate in the emerging technology.

Aviation came of age during World War I, but pilots remained all white. The army excluded blacks from flight classes based on the prevailing white belief that African Americans simply possessed neither the intelligence nor the physical skills to fly airplanes.

Even while whites held tenaciously to their old prejudices, one African American had already proved the fallacy of such beliefs even before the United States became actively engaged in World War I. A decade before the conflict, Eugene Jacques Bullard left the United States while a teenager to escape racism and discrimination. He immigrated first to Great Britain and then to France, where he trained as a boxer and fought all over Europe and the Middle East.

When war broke out in 1914, Bullard enlisted in the French Foreign Legion and suffered serious wounds while fighting as an infantryman. Upon his recovery, Bullard volunteered for flight training and in the summer of 1917 earned his pilot wings and promotion to sergeant. Some accounts say that in addition to his wings, Bullard received one thousand dollars from an American in France who bet that the black man could not learn to fly.

Known as the "Black Swallow," Bullard joined a French pursuit squadron whose ground crew painted his plane's fuselage with a heart pierced by an arrow and the motto All Blood Runs Red. When U.S. Air Corps squadrons joined the war, Bullard attempted to transfer into its ranks, volunteering to fly combat missions or to become one of the desperately needed flight instructors for the fledgling air schools back in the States. The U.S. military denied both requests. Bullard remained in his French squadron, where he soon scored a "probable" but unconfirmed kill of an enemy aircraft. On November 7, 1917, Bullard downed a German triplane fighter, witnessed by fellow pilots, for his second victory.

Unfortunately, his illustrious career as a fighter pilot ended after six months of flying combat missions when Sergeant Bullard became engaged in an altercation with a French officer over a racial comment. No details have survived except that after the incident Bullard returned to the infantry, where he remained for the rest of the war. Although Bullard came back to the United States during World War II, he never received credit for his pioneer efforts as a black aviator who definitely did know how to fly.

Despite the individual performances of Bullard, Stowers, and other brave soldiers and the black combat losses of 750 dead and more than five thousand wounded, African Americans in World War I achieved no recognition or equality for their heroic efforts. When the Allies marched down the streets of Paris in a victory parade on Bastille Day, July 14, 1919, both France and Great Britain had black soldiers representing their colonies in their ranks. Not a single black soldier marched with any of the U.S. units.

Because of the need, the SOS units remained in France to perform their usual duties to prepare equipment for transport and loading ships. In addition they undertook the dangerous tasks of clearing minefields and obstacles left on the many battlefields. The black SOS units also were given the unpleasant task of removing U.S. soldiers' remains from temporary burial places for reinterment in military cemeteries.

While the Ninety-second and Ninety-third Divisions were among the first combat units returned to the States, their early departure was not in recognition of their service; rather, it reflected continued resentment by their white officers about fraternization between black soldiers and white French women.

Discrimination continued against black soldiers regardless of their duties or performance. In early 1919 the Ninety-second Division prepared to board transports at Brest, France, to sail home. Company D, of the 367th Regiment, received orders at the embarkation station to assist in the loading of coal onto the USS *Virginia* and then remain aboard for transport home. Black infantrymen so excelled in the dirty, tiresome job of loading the fuel that U.S. Navy captain H. J. Ziegmine, commander of the *Virginia*, penned a letter of appreciation to the company commander. Ziegmine wrote, "I take great pleasure in commending you and the officers and men under your command in connection with the coaling of this ship and at the same time wish to express my appreciation of the good conduct and the high state of discipline of your command."

While Ziegmine recognized the conduct and discipline of the Company D soldiers in filling his coal bunkers, he had no intention of transporting those troops. He informed the embarkation office that "no colored troops had ever traveled on this ship" and that none ever would. The black soldiers, their letter of appreciation in hand, boarded a tug and returned to the Brest wharf, where, because of their late arrival, they had no place to sleep or anything to eat until the next day.

Once aboard ships which would accept them, the black soldiers and officers continued to suffer discrimination. They berthed in the least comfortable compartments and found areas of the ships restricted to whites only. When four hundred officers, all but one white, sailed home on the *Siboney*, whose mess could only seat two hundred, the captain directed three seatings for each meal: one for 200 white officers, one for 199 white officers, and a final seating for the single black captain. One white officer, Lt. Walter D. Binger, later wrote about the voyage, noting the dining arrangement and stating that to the best of his knowledge he was the only officer onboard to even speak to the black officer.

African-American soldiers aboard navy ships for transport home were unlikely to see many black sailors, if any at all. The policy of restricting black sailors to menial positions, such as messmen and servants to the officers, segregated them into limited areas of the ships. During the buildup for the war in 1917, U.S. senator Joseph S. Frelinghuysen had questioned the navy's policy of limiting shipboard positions for blacks. Secretary of the Navy Josephus Daniels replied by letter: "You are informed that there is no legal discrimination shown against colored men in the navy. As a matter of policy, however, and to avoid friction between the two races, it has been customary to enlist colored men in the various ratings of the messmen branch; that is cooks, stewards, and mess attendants, and in the lower ratings of the fireroom; thus permitting colored men to sleep and eat by themselves."

Daniels made no mention that, unlike in previous wars, the navy was having such little trouble filling its ranks with white enlistees that they made no effort to ease the restrictions on opportunities for black volunteers. As a result, although wartime expansion brought the total strength of the navy to 435,398, by June 30, 1918 only 5,328, or about 1.2 percent, were African Americans. The navy also continued its policy of not permitting blacks to earn commissions; the naval officer corps remained all white.

Black women also suffered discrimination in their efforts to support the war. In 1909 the National Association of Colored Graduate Nurses (NACGN) organized to improve educational and work opportunities for black women. When the United States declared war on Germany on April 6, 1917, Ada Thoms, cofounder of the NACGN, encouraged black nurses to join the American Red Cross, which assumed duties similar to that of the DAR during the Spanish-American War. The Red Cross registered interested African-American nurses but then rejected their applications on the grounds that the army did not accept black women.

Even when the Red Cross had difficulties meeting its goal of thirty-five thousand nurse volunteers, it continued to register but not accept blacks. It was not until the war had already concluded that black nurses could finally enter the Army Nurse Corps in December 1918—and again this improvement was due more to the result of necessity than a matter of equality. The postwar influenza epidemic that raged throughout the following year killed an estimated half million Americans, including military and medical personnel. In what it openly declared an "experiment," the army accepted eighteen black nurses and assigned them to hospitals at Camp Sherman, Ohio, and Camp Grant, Illinois.

Although their living, dining, and recreational facilities remained segregated, the nurses attended both black and white soldiers.

The only other opportunity during World War I for black women to directly assist those in uniform was with the Young Men's Christian Association (YMCA). Within the United States and overseas in France, the YMCA, authorized by the War Department, operated service centers to provide recreation, refreshments, and a comfortable place to write letters, read, or just relax. At the beginning of the war, the YMCA promised to provide equal services for whites and blacks, but like everything else within the military, these centers remained segregated.

More than two hundred black women served as hostesses in YMCA centers dedicated to African-American servicemen in the United States. Fifteen centers, operated by sixty-eight men and nineteen women, also served black soldiers in France during the war. In addition to the usual comforts provided by the YMCAs, those who served blacks also offered classes that taught thousands of previously illiterate or semiliterate black soldiers how to read and write.

The Allied victory in World War I left the United States in a position to join the ranks of world powers and to exert its influence around the globe. While the United States ultimately benefited by the war, African Americans did not. Once again blacks had to fight for the right to fight. And once again, after they had fought and many had died, their country neither recognized nor rewarded their service.

9

BETWEEN THE WARS: THE STRUGGLE CONTINUES

As a result of its participation in World War I, the United States gained international power and influence. While the country as a whole benefited from its new status, individual Americans—both black and white—had to adjust to the wide-ranging changes resulting from such a catastrophic event. Ironically, the war in Europe brought prosperity to black Americans; with peace, however, came a return of their former conflicts.

During the war itself, blacks in America benefited from increased job opportunities. Besides removing 4 million men from the labor pool and putting them in uniform, the war stopped the immigration of European whites that had traditionally filled jobs in factories, mills, and mines. To man industrial positions in the North left vacant by the draft and the lack of white immigrants, businesses turned to the South to recruit black workers. Between 1916 and 1918, an estimated 1 million blacks moved to the North for jobs and better opportunities—a movement, some declared, that produced individual rights for southern blacks that rivaled the end of slavery at the conclusion of the Civil War.

White America endured black workers in jobs previously held by whites out of economic necessity during the war and even tolerated black soldiers in Europe so that they could perform the dirty details of service and support. Many whites also endorsed black combat units in what they still considered "Sambo's right to die."

Once the war finally concluded, white Americans, wishing to revert to prewar employment practices that would allow returning white soldiers to regain their previous employment, expected blacks to accept their prewar subservient, less-than-equal status. Black Americans had other ideas, even though they were often powerless to implement them.

One initial problem for African Americans at the end of the war, beside the loss of job opportunities, was white American outrage, especially in the South, about the equality extended to black soldiers by the French and other Europeans, especially by French women, who treated them as equals rather than subhuman. Throughout the Deep South community leaders urged the formation of vigilante groups to prevent the wholesale ravishment of southern women. Sen. James K. Vardaman of Mississippi demanded that white southerners defend their wives and daughters against "French-women-ruined Negro soldiers." Mississippi governor Theodore G. Bilbo let it be known that his state would take no action against anyone responsible for lynching a black man accused of rape.

Vardaman, Bilbo, and other white leaders ignored the fact that during the entire time the Ninety-second Infantry Division served in Europe, only one of its black soldiers had been convicted of rape; the Ninety-third Division reported similar figures. Although rumors of sexual assaults by black soldiers in the service units had circulated in the American Expeditionary Force (AEF), these were also unfounded. In reviewing postwar support activities provided by the seventy-five thousand black soldiers, the judge advocate general of the Services of Supplies (SOS) reported only one incident of attempted rape and concluded, "The rape stories seem not to be substantiated."

Nevertheless, black soldiers returning home from the war met hostility, contempt, and violence from many whites, just as black veterans returning from Cuba had experienced. Instead of marching bands and grateful citizens to welcome them, black soldiers encountered mobs, complete with Ku Klux Klan members, who frequently beat them and stripped them of their uniforms. It was white America's effort to return African Americans to their "place."

One local city official, greeting a group of veterans returning to New Orleans, declared, "You niggers were wondering how you are going to be treated after the war. Well, I'll tell you, you are going to be treated exactly like you were before the war; this is a white man's country, and we expect to rule it."

The police rarely interfered with the mistreatment of black veterans and, in fact, on occasion participated in the mayhem. For some blacks reprisal was more severe than beatings or the humiliating loss of the uniforms in which they had risked their lives. In 1919 ten veterans were among seventy-seven blacks lynched by white mobs. A Georgia mob beat a black soldier, still in uniform, to death for merely violating a "whites only" Jim Crow law.

While White Americans reacted to black veterans and civilians as they had in other postwar eras, the Anglo majority was not prepared for the reaction of African Americans to the continued racism and discrimination. Blacks were unwilling to return to their "place," as defined by the white majority. Black veterans wanted to enjoy the freedoms and rights for which they had fought in Europe, and African-American civilians were determined to maintain the better jobs and pay they had earned in the wartime economy.

W. E. B. Du Bois, editorializing in the *Crisis*, continued to provide the most eloquent black voice as he explained the feeling of the African-American veterans in the May issue:

> We return from slavery of uniform which the world's madness demanded us to don to the freedom of civilian garb. We stand again to look America squarely in the face and call a spade a spade. We sing: This country of ours . . . is yet a shameful land. It lynches . . . steals . . . insults us. . . . We return from fighting. We return fighting. Make way for Democracy. We saved it in France, and by the Great Jehovah, we will save it in the U.S.A., or know the reason why.

Du Bois declared that his wartime objective of encouraging blacks to "close ranks" in support of the conflict had changed to a militant stance in pursuit of equality. In the same piece, Du Bois also wrote, "By the God in Heaven, we are cowards and jackasses if now that the war is over, we do not marshal every ounce . . . to fight a . . . more unbending battle against the forces of hell in our land."

When white enthusiasm to put blacks back in their prewar "place" collided with African-American passion for equality, the result was, of course, violence. During the last six months of 1919 more than twenty-five riots erupted across the United States when blacks and whites clashed, marking the time as the "Red Summer" for all the blood spilled. In Chicago the opposing races competed for jobs, housing, and recreation space. The Chicago riot in the summer of 1919 killed 23 blacks and injured 342, while white casualties totaled 15 dead and 178 injured.

Race riots in which blacks and whites fought in the streets during the Red Summer of 1919 occurred not only in large population centers like Chicago and Washington, D.C., but also in midsize cities like Omaha, Nebraska, and Knoxville, Tennessee, and in small towns like Longview, Texas, and Elaine, Arkansas.

Tensions persisted as whites blamed racial problems on foreign ideas brought home from Europe by black veterans. However, once whites realized that blacks, no longer docile in resisting oppression, were willing to arm themselves and to fight back, the riots waned, giving way to smaller-scale atrocities, such as lynching and burning of individual blacks by white mobs. The enforcement of Jim Crow laws symbolized a last-ditch effort to keep black Americans in their unequal status.

Black veterans and their families also experienced blatant official discrimination. When erecting monuments to the World War I dead, some communities omitted names of black casualties or placed them on separate markers. The American Legion permitted black veterans to join their organization, but only in newly formed, segregated posts. When the War Department organized a voyage to Europe for Gold Star Mothers of American soldiers buried in France, black women sailed on separate vessels.

The military made no effort to end discrimination. Blacks found themselves no more welcome as soldiers in the postwar army than they were as civilians in the towns and villages of white America. White army officers harbored racist agendas based on the belief that black subordinates had adversely influenced their careers during the war. These officers devalued the fighting abilities of African Americans and advanced the stereotype of the race's inferiority.

Neither the white commanders of the Ninety-second and Ninety-third Infantry Divisions nor senior white officers from other black units stood up for their soldiers. In postwar reports to the War Department and the Army War College, they stated that black soldiers were biologically and intellectually inferior and even added that black soldiers feared the dark and refused to fight at night. The white commanders reported that black officers failed to earn the respect of either their superiors or their black subordinates.

The senior officers placed all the responsibility for the poor performance by blacks in France on their African-American subordinate officers and enlisted men, completely ignoring the real causes: the poor training black units received, the low priority of supplies and equipment for black units, and the lack of quality white officers assigned to black units. According to the commander of the Ninety-second Division's 367th Infantry Regiment, "As fighting troops the Negro must be rated as second-class material; this is due primarily to his inferior intelligence and lack of mental and moral qualifications."

Charles Ballou, reduced in rank from major general to colonel in the postwar cutbacks in military strengths, also expressed doubts about the viability of black units. Ballou, bitter from his perception that the Ninety-second Division's soldiers had ruined his career, wrote to the assistant commandant of the Army Staff College on March 14, 1920, that the army should avoid creating large black units and send all black draftees to laborer and support organizations after induction. He declared that only after careful screening in the service units should the very best recruits be reassigned to combat regiments. Ballou's single recognition of black soldiers' capabilities came when he acknowledged a need for black officers at the lower levels, but again he emphasized the strict screening process they should undergo before receiving commissions.

Gen. Robert Lee Bullard also had nothing positive to say about the Ninety-second Division in his memoirs, published in 1925. In *Personalities and Reminiscences of the Great War*, Bullard wrote that black soldiers in his command proved to be reluctant to attack and that he "could not make them fight." Bullard declared black soldiers "hopelessly inferior" and concluded, "If you need combat soldiers, and especially if you need them in a hurry, don't put your time on Negroes."

Although there were some dissenting views in postwar studies about the capabilities of African-American soldiers, no one in the army stepped forward to challenge the reports or to champion black abilities. Some officers and War Department officials called for the total elimination of blacks in the army. They eventually reached a consensus to maintain a minimum number of black combat regiments, with the bulk of enlistees given service and labor assignments. Black units, regardless of mission, were to remain segregated and commanded by whites.

Those African-American officers who desired to remain in the post-war army and who applied for entry into the Regular Army received swift rejections, usually with no explanation. However, at least one black officer received notification that he was unqualified "by reason of qualities inherent in the Negro race." All black officers from the volunteer units received their discharges or reassignment to their state's National Guard.

When the National Defense Act of June 1920 declared that the future organization of the U.S. Army should contain only thirty thousand officers and men, all four veteran black regiments—the Ninth and Tenth Cavalry and the Twenty-fourth and Twenty-fifth Infantry Regiments—managed to survive the downsizing. Their survival sprang not from any great desire on the part of the War Department to keep them on the active rolls but on several factors based on perceived legalities and necessity.

The black community, and many within the military, interpreted the post–Civil War legislation that originally authorized the four regiments as providing a legal requirement for their retention on active duty. While this perception might not have withstood legal review, white officials were unwilling to take the issue to court and possibly set off additional protests and riots. Moreover, the regiments were needed for battalion rotations to the Philippines, a duty that had become traditionally met by the black cavalry and infantry units.

Also, for the army, the basic organization of combining two regiments into a brigade for combat necessitated the retention of two each of black cavalry and infantry regiments. Had one been eliminated, a black regiment would have had to combine with a white regiment to create an integrated brigade—an option which remained unacceptable to white America.

Another reason for the retention of the black units also dated back to the Civil War. Many whites, while desiring a segregated society and limited interaction with blacks, still supported the theory of "Sambo's right to be killed." If equality extended to nothing else, the right of the black man to die in combat remained unchallenged.

A study released by the army chief of staff in November 1922 supported the right of African Americans to fight and die in combat. The study simply and directly concluded: "To follow the policy of exemption of the Negro population of this country from combat means that the white population, upon which the future of the country depends, would suffer the brunt of loss, the Negro population none. . . ."

The chief of staff study of 1922 also emphasized the importance of maintaining white officers in the senior command positions but conceded that some black officers would have to be included because "it is not reasonable to expect that the Negro will be willing to serve in the ranks with no hope of a commission." Despite the study's findings, the officer ranks remained mostly off limits to blacks. By the time the army reached its reduced post–World War I strength, less than a half-dozen black officers remained in the Regular Army.

The primary source of Regular Army officers remained the U.S. Military Academy. However, after the graduation of Charles Young in 1889, West Point accepted no other black cadet until Alonzo Parham, from Illinois, entered the academy in 1929. Parham left after only a year, and in 1932, Benjamin O. Davis Jr., the son of one of the few black Regular Army officers, entered the academy. Davis graduated in 1936 and became West Point's fourth black graduate.

Few Americans, either white or black, paid much attention to the academy or the military as a whole in the years following the war. World War I became "the war to end all wars," and Americans focused on economic rather than defense issues. When the boom days of the 1920s collapsed with the Crash of 1929, the military experienced further budget cuts and personnel reductions. During the Great Depression of the 1930s, opportunities for all Americans remained limited.

The numbers of blacks in the U.S. Army and their degree of equality continued to diminish. Although the four black regiments remained on active duty, their soldiers rarely trained for combat or operated as complete units. Instead, black cavalrymen and infantrymen were detailed to post maintenance and support positions—grooming the grounds, cleaning stables, and other laborer jobs. The Tenth Cavalry, stationed at West Point for much of the decade, acted as orderlies and servants for the academy's staff and cadets. At Fort Benning, Georgia, soldiers of the Twenty-fourth Regiment rarely practiced rifle marksmanship or field tactics. Rather, they acted as vehicle drivers, cooks, and stable hands in support of the post's white units.

In 1933, President Franklin Roosevelt formed the Civilian Conservation Corps (CCC) to fight unemployment by providing young men jobs. Although the law authorizing the CCC clearly specified that its

Capt. Charles Young, the third black graduate of West Point, in Ninth Cavalry uniform (ca. 1905).
(U.S. ARMY MILITARY HISTORY INSTITUTE)

ranks were open to all, regardless of race, and that everyone would receive equal treatment, the CCC accepted proportionately fewer blacks than whites and, after the first few months, segregated those few into their own camps.

Army officers from the regular ranks and reservists called to active duty administered the CCC. White officers filled the leadership positions at both the white and black camps. During the program's first two years, not a single black reserve officer received orders for CCC duty. Not until Emmett Scott, former adviser to President Wilson during World War I and now president of Howard University, and NAACP officials complained to Roosevelt did the War Department finally agree to activate black reserve officers. Even this improvement proved to be a token effort, with only chaplains and doctors called up.

Their numbers never exceeded twenty on active duty at any time for the duration of the CCC.

During the post–World War I period the African-American press continued to be the greatest ally and supporter of black participation in the military. While their austere budgets during the actual fighting in France had limited the number of reporters sent to the war zone, black newspapers closely followed the return of the veterans after the armistice and developments during the 1920s and 1930s. African-American newspapers published stories about the assignment of black combat troops to garrison support positions and described the separate and mostly unequal eating and recreational facilities on the various posts. The African-American press also lobbied for the opening of military specialties closed to black soldiers, including the newly formed air corps, and editorialized in favor of a peacetime black division.

These efforts produced little. Black soldiers in the combat regiments continued to perform garrison labor and support functions, and the number of replacements did not match discharges. By the late 1930s less than four thousand African Americans were on active duty, fewer than had been in the ranks prior to World War I. In response, protests, again led by the black press, demanded that the army accept black volunteers in proportion to their population. In 1938 the army finally agreed to accept black volunteers according to their population ratio of 9 to 10 percent. Despite the official policy, there was no immediate, substantial increase in the induction of black recruits.

Blacks seeking to serve in the National Guard faced even greater difficulties. The War Department left untouched the policy of leaving the entrance requirements and racial composition of the National Guard to each individual state. All thirty of the National Guard regiments that formed its sixteen authorized divisions were all white. The only black combat regiments in the National Guard were the New York 369th, the Eighth Illinois, and the 372nd Regiment, composed of far-flung battalions and companies from Maryland, Massachusetts, New Jersey, Ohio, and the District of Columbia. By 1939 blacks totaled only about 2 percent of the strength of the Regular Army and National Guard.

While the army provided only a few positions and limited opportunity for African Americans in the post–World War I era, the other U.S. armed services offered even less. The Marine Corps continued to exclude blacks and even refused to employ African-American civilians as messengers in its Washington, D.C., headquarters. From the end of World War I to 1932, the U.S. Navy accepted few black enlistees and limited them to servant and laborer positions. The few black gunner's mates and other "on deck" sailors remained on active duty until they completed their enlistments or retired; whites then replaced them in those positions. By 1932, of the 81,000 men in the navy, only 441 were black—the lowest total in American history.

Only when necessity reopened navy enlistments to African Americans did the numbers of blacks increase. In 1932 the planned granting of independence to the Philippine Islands reduced the number of Filipino volunteers, forcing the navy to recruit blacks to fill vacant messmen positions. Naval officers, fond of the subservient Filipinos, fretted over their loss and expressed their desire to recruiters to seek only black volunteers from the South who were accustomed to white control and were not as educated or independent as those from the North.

So the ranks of African Americans in the navy once again grew, reaching a total of more than twenty-eight hundred by mid-1939, but the vast majority remained servants and mess attendants. When black leaders and a few congressmen complained about the recruitment limitations, a committee of senior navy officers and department officials defended the policy, stating that "the enlistment of Negroes (other than as messmen) leads to disruptive and undermining conditions."

Limitations on black enlistees extended beyond their assignment as mess attendants. Although both black and white recruits received twenty-one dollars per month after their enlistment, blacks, rated as mess attendants third-class, were not eligible for promotion or the resultant pay raise during their first year. On the other hand, a white recruit could be promoted every three months and by the end of a year could become a petty officer receiving a monthly pay of fifty-four dollars.

The navy also maintained its all-white officers' corps. In an active navy that accepted African Americans only as servants in the mess and as help in the kitchen, no opportunity existed for a black sailor to pursue a direct commission. The only other avenue to achieve officer rank remained the U.S. Naval Academy, but it had not accepted a black appointee since 1873 and had never graduated a black midshipman.

Many black leaders felt that the mission of the academy—to develop leaders with a "potential for future development in mind and character to assume the highest responsibilities of command, citizenship, and government"—conformed exactly to what they wanted for young African Americans. Finally, their lobbying once again opened Annapolis to black midshipmen, but only briefly and with negative results.

In 1936, James Johnson of Illinois received an appointment but remained at the academy for only eight months before resigning because of medical problems. The following year, George Trivers reported for classes but left after a month because of poor marks in deportment and English. Both black men suffered from severe hazing from the white midshipmen, and at least some of the white faculty discriminated against them in the classroom.

By the end of the 1930s some Americans realized that world war was threatening the peace—indeed, the security of the United States. A few understood that America would once again have to call upon blacks to preserve freedoms they could not enjoy equally. In March 1939 an editorial appeared in the *Crisis* that detailed the problems with Jim Crow policies in the military:

> Judging from prevailing Jim Crow practices in the armed forces of the United States today, the next war . . . will see the same gross maltreatment of the Negro soldier seen in the World War. For today Negroes are banned from the newer arms of service, including aviation and other branches; Negroes may serve in the U.S. Navy only as menials . . . Negro regular army soldiers are kept out of active service. . . . These are but a few of the many discriminations.

The *Crisis* article concluded that these discriminatory practices not only prevented equality but also endangered the security of the country. It warned, "Such a policy is contrary, not only to decency, but to the security and well-being of the American government." World War II would prove those views correct.

10

WORLD WAR II: ARMY

At the end of the 1930s black Americans lived in a segregated society governed by Jim Crow laws that prevented their enjoying the full benefits of the free society they had helped to form and maintain. Because they had been able to make few inroads into the white power structure of the civilian community and because their rights and numbers in the military continued to erode, African Americans began to concentrate their efforts to gain equality through internal unity. Black community leaders, labor organizers, and newspapers took on the role of activists to garner political power for African Americans, who now represented 10 percent of the country's population and tended to vote as a group at the polls. Election years became the greatest ally to black progress, and African Americans began to exploit their emerging advantage.

Racial pride was also on the increase. More blacks attended colleges and universities, and the race experienced a cultural renascence that produced entertainers and artists who appealed to both the black and white communities. Athletes, such as boxer Joe Louis and track star Jesse Owens, proved that when given the opportunity, blacks could excel.

While athletics and the arts increased the awareness of black skills and enhanced racial pride in the African-American community, real power resided with labor. In 1936 the American Federation of Labor granted full membership to the Brotherhood of Sleeping Car Porters, under the leadership of A. Philip Randolph, thus making it the most powerful black labor organization in the country.

By the end of the 1930s, African-American organizations and the black press were ready to demand equality and justice. The NAACP and the National Urban League campaigned for black rights, and black newspapers, particularly the *Atlanta Daily World*, the *Chicago Defender*, the *Norfolk Journal and Guide*, and the *Pittsburgh Courier*, exposed mistreatment and discrimination.

Although often kept away from the polls in the South by poll taxes and literacy requirements, black voters presented a formidable force. Their switch from supporting Republican candidates, a tradition dating back to Lincoln and the Civil War, to voting Democratic in 1932 helped put Franklin D. Roosevelt in the White House. Although Roosevelt supported black equality only when it was politically expedient, his wife, Eleanor, became a strong advocate for black rights.

With the prospect of war in Europe and the Pacific looming on the horizon, blacks found a new sense of power when their emerging political clout coincided with the military's traditional need of black manpower in times of conflict. While World War II would provide some legitimate progress toward equality, the diehard racism of whites would ultimately yield only token gains and limited opportunities.

Although they had learned during the past half century that loyal service in time of war did not guarantee racial progress when peace returned, African Americans nevertheless believed that serving in the military in war would provide personal and general opportunities for their race. Young blacks viewed the military in the same way as had their fathers and grandfathers. Service in uniform offered adventure, upward mobility, and improved status. Blacks as a whole also clung to the idea that if they again fought for their country, surely they would reap the benefits. While some blacks had doubts based on past history about such possible gains, most thought that the magnitude of the impending war and the number of blacks prepared to serve would surely make a difference this time.

Opposition to military service in the African-American community centered on concerns over blacks fighting foreign enemies when discrimination still thrived at home. However, as they learned more about Hitler and Germany's beliefs in a "master race" that considered blacks as genetically inferior, most of the resistors rallied to endorse the war effort.

Most African Americans willingly supported the prewar buildup of the armed forces, but they understood that no true advances in equality could take place in segregated forces. Initially, the black community demanded complete equality in the armed forces through total integration. Charles H. Houston voiced the feelings of the black community in a letter to President Roosevelt on October 8, 1937. Houston, a veteran of World War I and a special counsel to the NAACP, requested that the president "give Negro citizens the same right to serve their country as any other citizen, and on the same basis."

In another letter dated the same day, Houston wrote Secretary of War H. A. Woodring, assuring him that the black community would remain loyal in the impending war but warning that "the Negro population will not silently suffer the discrimination and abuse which were heaped upon Negro soldiers and officers in World War I. We urge you to remove all racial barriers to service in all branches of the army."

At the time of his writings, Houston was unaware that the War Department had already conducted a study earlier that year to establish preliminary policies for the treatment of blacks in the event of war. The resulting plan called for equality in the numbers of servicemen but maintained all the vestiges of segregation. Specifically, the policy lifted the restrictions on the number of blacks to be inducted in order to bring the percentage of black soldiers in the army in proportion with their total population. The plan also called for these men to be mobilized before an actual national emergency or the declaration of war occurred. Lower-ranking officer positions in black units were to be reserved for African Americans, but at no time would black officers command white troops. The plan included additional authorization for white officers to be trained to command and serve in senior positions in black units.

The 1937 War Department plan for blacks in the next war was not formally classified, nor were its contents made public. Knowing that black leaders and their white supporters would object to the plan's segregation policy, the War Department avoided questions or arguments by merely not sharing the information.

In the meantime, black organizations and newspapers demanded increased black participation in the military. Finally, in 1939, the army began accepting greater numbers of blacks to work toward the 1937 enlistment goal of 10 percent. To absorb the new troops that year, the army formed the Forty-seventh and Forty-eighth Quartermaster Regiments as all-black units and during 1940 added an artillery and an engineer regiment as well as smaller coastal artillery, transportation, and chemical units.

While the black community continued to believe that equal opportunity in the military was paramount to gaining true equality, white military leaders still looked upon African Americans as inferior and ill qualified to be soldiers or sailors. An Army War College committee that studied the capabilities of black soldiers in 1939–40 concluded that the African American was "far below the white in capacity to absorb instruction." The committee members included a story to illustrate their evaluations of the capabilities of black officers, a story which, in itself, condemns the views of the evaluators, not the capabilities of the evaluated. The account went like this:

> A staff officer stopped at a crossroads and asked a Negro soldier which way one road went.
> "I don't know, sir."
> "Where does that road go?"
> "I don't know, sir."
> "Well, what are you here for?"
> "I don't know, sir."
> "Who put you here?"
> "The Captain, sir."
> "Where is the Captain?"
> "The Captain? He's right over here, sir, but he won't help you none. He's a nigger too!"

Black leaders lobbied for both an increase in the number of African Americans in the armed forces and their total integration into all aspects of military service. The immediate result was antidiscrimination terminology in important pieces of legislation. However, it lacked enforcement authority to end segregation within the military or to increase the number of blacks in uniform.

On September 14, 1940, Congress passed the Burke-Wadsworth bill, which authorized the first peacetime draft in American history. Two days later, President Roosevelt signed it into law as the Selective Service and Training Act of 1940. The main provision of the act called for the registration of all men between the ages of twenty-one and thirty-five and the induction of 800,000 draftees. As a concession to black leaders, Section 4(a) of the law stated: "There shall be no discrimination against any person on account of race or color."

Although it promised no "discrimination," the act did not provide any requirement or authority to eliminate segregation. In fact, Section 3(a) of the same act provided army and navy authorities "unlimited discretion" in deciding who to accept into their ranks and how to employ those whom they accepted.

The provisions of the Selective Service Act did little to quell the concerns of the black community about military segregation. On September 27, Walter White of the NAACP, A. Philip Randolph of the Sleeping Car Porters, and T. Arnold Hill of the National Youth Administration met with Secretary of the Navy Frank Knox and Assistant Secretary of War Robert Patterson to present their ideas on eliminating discrimination within the services. The seven-point program presented by the African-American leaders centered on integration of black officers and enlisted men throughout the armed forces, with skill positions and rank limited only by their personal abilities.

The demands were not well received. When newly appointed Secretary of War Henry L. Stimson was briefed on the meeting, he noted in his diary that he did not support the seven points: "I saw the same thing happen twenty-three years ago when Woodrow Wilson yielded to the same sort of demand and appointed colored officers to several of the Divisions that went over to France, and the poor fellows made perfect fools of themselves and one at least of the Divisions behaved very badly. The others were turned into labor battalions."

On October 9, Stephen Early, the White House press secretary, issued a statement initiated and approved by President Roosevelt and prepared by the War Department to define the army's policy "in respect to Negro participation." Early announced that the number of blacks inducted into the army would be in proportion to their population numbers and that all combat and noncombat units, including aviation, would accept black soldiers. Officer candidate school (OCS) would be open to blacks, but graduates could serve only in black units where white officers would hold command positions. The most important part of the statement, and the most disappointing to black Americans, was the affirmation of the standing policy of "not integrating colored and white enlisted personnel in the same regimental organizations."

The policy justified segregation by stating that it had proved satisfactory "over a long period of years, and changes would produce situations destructive to morale and detrimental to the preparations of national defense."

The contents of the policy alone were distasteful enough to upset the black community, but the situation worsened when Early implied that White, Randolph, and Hill had conferred with the president instead of just the War and Navy Department representatives and that they supported the plan. The three black leaders immediately protested the misrepresentation, and newspaper editorials railed against the policy. Then, to make matters worse, a few days later, Early kicked a black New York policemen in the groin when the law officer refused to let the press secretary cross a barrier protecting the president.

Early apologized to the policeman and to the New York police commissioner and wrote a letter of apology to White for misrepresenting the black leaders' opinions about the announced policy. Roosevelt, in the midst of campaigning for an unprecedented third term, extended his regrets about Early's errors to black leaders as he intensified his campaign to counter the damages and to prevent losing the black vote in the November election. Black political power was indeed beginning to match the influence of the wartime need for black soldiers and sailors in gaining concessions for African Americans.

As a means of damage control, Roosevelt communicated with black leaders to assure them that he supported the admission of blacks into all military specialties and promised that blacks would have an equal opportunity to gain commissions as officers, leading to "appropriate commands." On October 25, a week before the election, Roosevelt announced the promotion of Col. Benjamin O. Davis Sr., the father of the fourth black West Point graduate, to the rank of brigadier general—the first African-American flag officer in the U.S. military. On November 1, Roosevelt appointed Judge William H. Hastie, dean of the Howard University Law School, assistant secretary in the War Department to advise on black military issues. The White House also ensured that Gen. Lewis B. Hershey, the national director of the Selective Service System, announced prior to the election the appointment of a black officer, Maj. Campbell C. Johnson, to his staff to advise and assist in questions about the draft.

Brig. Gen. Benjamin O. Davis Sr. and Capt. Benjamin O. Davis Jr. (ca. 1943).
(U.S. ARMY MILITARY HISTORY INSTITUTE)

Roosevelt succeeded in retaining the black vote in his election to a third term. However, even though blacks supported the president in the election, most saw his damage-control measures for what they were and understood that the primary issue of segregation remained unchanged. The *Baltimore Afro-American* labeled Roosevelt's actions as "appeasement," and the *Guild Lawyer* declared: "The promotion of Davis . . . is another incident of our traditional practice to single out an individual for honors, at the same time to keep the mass of Negroes in inferior status or suppressed."

On the day Davis advanced to brigadier general, Secretary of War Stimson noted in his diary his belief that the promotion was of political rather than substantive value. Stimson sarcastically wrote, "I had a good deal of fun with Knox (secretary of the navy) over the necessity that he was facing appointing a colored admiral and a battle fleet full of colored sailors. . . . and I told him that when I called next time at the Navy Department with my colored brigadier general, I expected to be met with the colored admiral."

On the other hand, Judge Hastie did not see his own appointment as an assistant secretary in the War Department as a "promotion" and initially rejected the offer because it was largely a bureaucratic position with little or no power to actually make change. In his view, Emmett Scott, who had filled a similar position during World War I, had the authority only to respond to complaints from black soldiers, without the power to advance justice or social change within the military. Hastie finally accepted the position, but only on the condition that the War Department acknowledge the judge's strong opposition to a segregated military. After assuming his post, Hastie released a statement declaring: "I have always been constantly opposed to any policy of discrimination and segregation in the Armed Forces of this country. I am assuming this post in the hope that I will be able to work effectively toward the integration of the Negro into the army and to facilitate his placement, training, and promotion."

While integration of the military remained Hastie's primary objective, he conceded that Roosevelt had met many of the black community's demands and they would now have to respond in kind to continue their advancement. In addition to his post-appointment stance on integration, Hastie also declared:

> The man in uniform must grit his teeth, square his shoulders, and do his best as a soldier, confident that there are millions of Americans outside the armed forces, and more persons than he knows in high places within the military establishment, who will never cease fighting to remove all social barriers and every humiliating practice which now confronts him. But only by being, at all times, a first-rate soldier can the man in uniform help in this battle which shall be fought and won.

In the months following Roosevelt's reelection, black leaders recognized that other than the token appointments, the administration was not honoring its promises of opening more specialties to greater numbers of African-American volunteers and draftees. While the census of 1940 revealed that blacks composed 9.8 percent of the total U.S. population, fewer than 1 percent of appointees to the staffs of local draft boards were African Americans. States in the Deep South had few blacks on their draft boards, and these perfunctory appointees had no authority in decisions concerning which whites the boards would select for duty. By November 1941 blacks made up only 5.9 percent of the army's total rather than the nearly 10 percent promised by legislation and by the president. Writer and poet Langston Hughes noted the irony that while whites did not want blacks in military dress, they preferred them in other kinds of uniforms. Hughes wrote, "We are elevator boys, janitors, red caps, maids—a race in uniform."

Unhappy with the lack of progress following the preelection appeasement appointments, including the foot dragging in drafting blacks, A. Philip Randolph, in January 1941, called for a march on Washington the following July 1 to protest the exclusion of African Americans "from the defense industries and their humiliation in the armed forces." As the leader of the Brotherhood of Sleeping Car Porters, Randolph's primary interests were equal job opportunities for blacks in the rapidly expanding war industries, but his ideas about integration also extended into the military. In its annual meeting, the NAACP endorsed Randolph's movement and promised to join the demonstration. Randolph's threat of 50,000 to 100,000 blacks and supporters marching on Washington worried government officials that the protest might appear to Tokyo and Berlin as a weakness in American resolve and unity to defend itself.

Eleanor Roosevelt met with Randolph, asking him to cancel the march, and finally the president himself met with the labor leader to seek a resolution. Roosevelt agreed to make some concessions, and although some black groups encouraged Randolph not to settle for anything less than total integration of the workplace and the military, the labor leader finally agreed to call off the march. On June 25, 1941, only a week before the planned demonstration, Roosevelt signed Executive Order 8802, the Fair Employment Practices order, which banned discrimination in defense industries and the government.

While the military remained segregated and Jim Crow discrimination laws still dictated a lifestyle far from equitable for black Americans, the country had its first official employment nondiscrimination policy: "It is the duty of employers and labor organizations to provide for the full and equitable participation of all workers in the defense industries without discrimination. . . . All departments and agencies of the Government of the United States concerned with the vocational and training programs for defense production shall take special measures appropriate to assure that such programs are administered without discrimination."

The Fair Employment Practices order provided a milestone in the struggle of America's black community to gain equality for all and it turned the focus of the movement toward emphasizing the unacceptability of segregation in a democratic society. However, the actual order had few provisions for enforcement, and the opening of employment opportunities to blacks became more a result of need than any effort by the government or business or industry to provide equal opportunity.

Even as more Americans realized the hypocrisy of maintaining segregation while preparing to fight a war to defend the freedoms of others, discrimination against blacks remained a fact of daily life. Separate facilities for blacks and whites continued to exist throughout the South and other parts of the country, and construction plans even called for segregated black and white air-raid shelters in Washington, D.C.

At times the separation of the races escalated from the hypocritical to the absurd. In response to requests from the army and the navy, the American Red Cross announced in November 1941 that it would no longer accept blood donated by blacks because the "white men in the service would refuse blood plasma if they knew it came from Negro veins."

The absurdity in the announcement lay in the fact that it had been a black man, Charles Drew, chief surgeon at Washington's Freedmen's Hospital and a pioneer blood researcher for the Red Cross, who had developed the procedure for extracting plasma from whole blood. Drew resigned from his position with the Red Cross when the organization made its racist announcement.

While black organizations and leaders still protested separate facilities and humiliations like the refusal of their blood, they had at least secured equality in the workplace on paper. Now they concentrated their efforts on ending segregation in the armed forces. Judge Hastie lobbied from his position in the War Department for total integration of the army. On December 1, 1941, the senior officer in the U.S. Army, Gen. George C. Marshall, responded to a formal request from Hastie to integrate the force by stating, "The settlement of vexing racial problems cannot be permitted to complicate the tremendous task of the War Department."

The bombing of Pearl Harbor and the declaration of war by the United States the next week provided the army a ready excuse to continue to ignore racial "complications." Two days after the United States declared war, a spokesman for the adjutant general's department made a statement summarizing the army's position on integration that would, in one form or another, remain the War Department's stance for years into the future: "The army did not create the problem. . . . The army is made up of individual citizens . . . who have pronounced views with respect to the Negro. Military orders will not change their views. . . . The army is not a sociological laboratory."

Once the fighting actually began, neither white political leaders nor military commanders were interested in "sociological" experiments or equality for African Americans. The small prewar advances blacks had made evaporated when the United States became actively engaged in World War II. The white power structure treated blacks in the military as it had in World War I, acting as if advances against racism and discrimination had never occurred. African Americans, who had to fight for the right to fight while resisting menial support positions, organized into segregated units commanded by white officers.

Despite continued discrimination, African Americans came forward, as in all previous wars, to share in their country's defense and to willingly risk their lives for freedoms that they did not share. However, this time, blacks preparing to fight in World War II had fewer illusions that their service and sacrifice would yield civil rights and equality. They were willing to fight, but they had learned that any true advances in their social standing would come not from their military service but from racial unity and political influence. The desire of politicians to win the black vote in the two- and four-year election system of the United States would prove to be a greater ally of equality than the service of hundreds of thousands of black men and women in uniform.

During the war the American black press provided the voice of unity and pride, encouraging blacks to stand up for their rights at home and in the military. Around the world a "V," formed by the index and middle finger, symbolized "victory" for the Allies. Only a few months into the war the *Pittsburgh Courier* initiated the Double V campaign

Activation ceremony of the Ninety-second Infantry Division at Fort McClellan, Alabama, on October 15, 1942. (U.S. ARMY MILITARY HISTORY INSTITUTE)

to encourage blacks to hold up V's with both hands to symbolize their efforts to achieve victory at home against discrimination and victory abroad against the Axis.

The American political and military leaders, little interested in black gains at home, were well aware that their service was essential to victory abroad. However, racism in local draft boards impeded their entering the service. Those drafted or allowed to volunteer found the military ill prepared to receive them.

Draft boards, particularly in the South, often placed black men at the end of their eligible lists. When finally drafted, blacks faced further discrimination at the reception stations, where white officials, using the justification that blacks did not possess the intelligence to master sophisticated weapons, assigned the majority to labor and support units rather than combat regiments.

Because of limited educational opportunities, 80 percent of black draftees scored in the bottom two-fifths, known as Categories IV and V, in their induction testing, scores which denied them entry into the service. When the need for draftees exceeded available qualified men in 1943, the army finally initiated special training units to educate previously rejected recruits to a passable level. More than 136,000 black soldiers and 160,000 whites attended twelve-week special schools, with most advancing their scores sufficiently to meet the draft requirement. In fact, the black remedial students proved that with equal educational opportunities they could match or exceed the success of the white students. Black men completed the remedial courses at a rate of 85.1 percent, compared to 81.7 percent of whites.

Blacks accepted to active duty faced rigid segregation at the training centers. All barracks, dining facilities, recreation places, and even hospitals remained segregated. On the same post, two of everything had to be prepared to maintain the separation of the races during training and off-duty hours. These facilities were separate but far from equal. Black trainees and units lived in the oldest, poorest-maintained facilities, located in the camp's most remote areas. One of the few buildings the races shared was the post chapel, and even then some of the interdenominational houses of worship posted religious service schedules for "Protestants," "Catholics," "Jews," and "Negroes."

Because most of the training camps were located in the South, black soldiers found themselves even less welcome in the neighboring towns than on "whites only" sections of their posts. Recreational facilities off base remained just as segregated as those on the post—if available at all. Jim Crow laws forcing blacks to the "back of the bus" and excluding them from restaurants and theaters remained. Many black soldiers found local towns so hostile that they remained in camp, feeling imprisoned and with no access to the outside world. Even staying in camp was not always safe. For no known reason other than the color of his skin, one black soldier, Prvt. Felix Hall, was lynched at Fort Benning, Georgia, by murderers the army never identified or punished.

After an inspection trip to army camps throughout the United States in 1943, Brig. Gen. Benjamin Davis noted in a memorandum:

> There is still great dissatisfaction on the part of colored people and soldiers. They feel that, regardless of how much they strive to meet War Department requirements, there is no change in the attitude of the War Department. The colored officers and soldiers feel that they are denied the protection and awards that ordinarily result from good behavior and proper performance of duty. . . . The colored man in uniform receives nothing but hostility from the community officials. . . . The colored man in uniform is expected by the War Department to develop a high morale in a community that offers him nothing but humiliation and mistreatment . . . Officers of the War Department General Staff have refused to attempt any remedial action to eliminate Jim Crow. In fact, the army, by its directions and by actions of commanding officers, has introduced Jim Crow practices in areas, both at home and abroad, where they have not hitherto been practiced. . . ."

More than 2.5 million African Americans registered for the selective service during World War II. Despite the discriminatory draft, nearly half of this number were inducted into the armed forces, about 75 percent of whom went into the army. While the percentage of blacks in the army increased from 5.9 percent at the time of Pearl Harbor to a high of 8.7 percent in late 1944, African-American soldiers in the army never at any time during the war reached the nearly 10 percent ratio of the total population, as promised by Roosevelt and the War Department.

Through special training and by their performance once in uniform, blacks could master any skill in the army, yet most were still assigned to laborer and service units. Only about 15 percent of black soldiers secured combat assignments. As they had in World War I, black support troops unloaded ships and moved supplies and ammunition to the frontline troops.

During the war African Americans also served in engineer units that built roads and bridges around the world to ensure the movement of military cargo to the front. Blacks composed one-third of the engineers who built the Alaskan Highway and 60 percent of the troops that constructed the Ledo Road across extremely difficult terrain from Burma into China.

Transportation companies of African-American truck drivers, mechanics, and administrative personnel made up 75 percent of the famed Red Ball Express, which moved supplies, fuel, and munitions from ports along the French coast to the rapidly moving American armor and infantry units during their race across Europe in the war's final months.

While black support units were deployed to both the European and Pacific theaters early in the conflict, black combat units did not engage in action until late in the war. In 1942 the army reactivated the two black infantry divisions, the Ninety-second and Ninety-third, which had fought in World War I, and formed a new Second Cavalry Division, which contained the Ninth and Tenth Cavalry Regiments.

Following their activation, the black divisions spent extensive time training and preparing for combat. However, as the months passed, white divisions, organized at later dates, were deployed overseas, while the black divisions remained in the United States. Within the black units and in the black community, many believed that the War Department was deliberately depriving the divisions from gaining the opportunity to prove themselves in combat and to earn honor on the battlefield.

Once again need figured in the equation that finally authorized the deployment of the black divisions overseas, but politics proved to be even more influential. Increased Allied operations in both the European and Pacific theaters required additional combat divisions regardless of their color, but it was not until the election of 1944 neared that the War Department began plans to actually send the black divisions overseas. President Roosevelt, needing the black vote for his fourth term and badgered by the black press and leaders, finally pressured the army to use the black divisions for their intended purpose.

Although the African-American community stood mostly united in its support of the war against the Germans and their "master race" theory, many blacks felt some empathy for the Japanese, whom they considered another minority group often exploited by white-dominated countries. Nevertheless, blacks joined the army ready to stand against all enemies of the United States, and it was the Japanese and not the Germans whom they were first sent to fight.

The Ninety-third Infantry Division was deployed from San Francisco to the Pacific theater during the first two months of 1944. Even though the division contained the veteran Twenty-fourth and Twenty-fifth Infantry Regiments, which had served on active duty for more than seventy-five years, the soldiers in it never fought as a unit and saw little combat. Upon arrival in the Pacific theater, the Twenty-fifth Regiment served on Guadalcanal in the Southern Solomons; the 368th, on Banika in the Russells; and the 369th, on New Georgia in the Central Solomons. Fighting for these islands had concluded before their arrival, and the three black regiments performed security missions and acted as laborers, unloading ships and cleaning up battle damage.

Complaints from black leaders in the States about the employment of the combat regiments as laborers finally brought orders to commit the African Americans to combat. On March 11 the First Battalion, Twenty-fourth Infantry, joined the white Thirty-seventh Infantry Division against the Japanese on the island of Bougainville. During their first night they turned back a Japanese attack and in the following days conducted an attack of their own. Although black soldiers showed inexperience in not always controlling their fire and their officers had difficulty reading their maps in the thick jungle, the battalion performed satisfactorily.

On March 28 all three battalions of the Twenty-fifth Infantry Regiment, attached to the Americal Division, joined the battle for Bougainville. Early in the fight, experienced and battle-hardened Japanese troops ambushed Company K of the Regiment's Third Battalion, killing ten and wounding twice that number. The remainder of the company hastily withdrew in an unorganized fashion that spawned rumors in adjacent white units that the entire regiment had broken under fire.

The reactions of Company K were not unusual for units new to the jungle and facing veteran enemy troops for the first time. New white units experienced the same disorganization on Bougainville and during other island campaigns, but unlike the black Twenty-fifth Regiment, no one spread rumors about their breaking and running in combat. Actually, Company K and the rest of the regiment performed well during the remainder of the campaign and helped defeat the Japanese. Several of the black soldiers earned valor decorations.

Even though the Twenty-fifth, "bloodied" on Bougainville, were now veterans themselves, they did not participate in further island-hopping offensives. Along with the other regiments of the Ninety-third Division, they served as occupation troops on captured islands and worked mostly as laborers. Their only real action came during the occasional patrol to capture or eliminate the few remaining Japanese who had hidden in the jungle rather than surrender. A Twenty-fifth Regiment patrol on the island of Morotai captured Col. Kisou Ouchi, one of the highest-ranking Japanese prisoners of war.

Only the Twenty-fourth Regiment saw any sustained action in the final year of the war; they fought pockets of resistance on Saipan and Tinian, in the Mariana Islands, from May 1945 until the end of the conflict. The individual soldiers of the regiment earned their combat infantryman badges, and the unit as a whole received a campaign-ribbon battle star.

In addition to the infantry regiments being mostly withheld from combat, black support units, in which more than 80 percent of African Americans in the Pacific served, also faced discrimination. Blacks assigned to support units stationed in Australia complained about segregation on military installations and often engaged in fights with white American troops over the use of Australian facilities. After a large racial clash in Brisbane between black and white Americans in 1943, Col. C. H. Barnwell Jr., the inspector general of the U.S. armed forces in Australia, investigated the matter and wrote his commander requesting "that recommendation be made to the War Department that no additional Negro troops be sent to Australia; that Negro troops (already in country) be stationed in northern and northeastern Australia, where some association may be had with the Australian Negro, commonly known as an aborigine."

All across the Pacific theater black combat and support units served in some of the more remote locations, where their contributions to the victory over Japan went mostly unnoticed and unrewarded. While they did not partake equally in the honor and glory of the Japanese defeat, black soldiers certainly received more than their share of attention when they did anything wrong.

African-American soldiers committed no more or no fewer military and civil crimes on a proportional basis than did whites. Their punishment, however, was more severe at all levels, including execution for capital crimes. Of twenty-one American soldiers executed in the Pacific—with the final approval of Gen. Douglas MacArthur—during and immediately after the war, eighteen were black.

The army hung six of the eighteen blacks in the Pacific theater on October 2, 1944, for the gang rape of two white American nurses on New Guinea. The military-police investigation of the crime had many shortcomings, and the accused soldiers did not receive adequate representation at their trail. The court-martial board ignored evidence showing that at least one, and possibly two, were innocent and found all six members of the 808th Quartermaster Amphibian Truck Company guilty. In his last letter home to his mother, Pvt. Lloyd L. White wrote: "It is difficult to write this letter to you. I have to tell you that the Military Courts have pronounced a very heavy sentence upon us, for which we are not rightly guilty of. Maybe the answer lies in it because of the color of our skin. . . ."

In the Mediterranean and European theaters, black support units provided the logistic functions for frontline white divisions. The Second Cavalry Division, containing the veteran Ninth and Tenth Cavalry Regiments, sailed for the Mediterranean in the spring of 1944. Although organized, trained, and prepared to join the Italian campaign, the Second Cavalry disbanded on May 10, 1944, its soldiers reassigned to support units in North Africa. The War Department explained the action by stating that a need for additional support troops existed and that the cavalrymen were trained for an obsolete function of no use in modern warfare. The explanation made no mention of the transition of white cavalry units to mechanized warfare or to the infantry.

Relentless complaints from black leaders about the failure of the War Department to deploy black combat troops to the front finally led to the commitment of the Ninety-second Infantry Division to the Italian campaign of 1944. The Ninety-second, activated at Fort McClellan, Alabama, on October 15, 1942, conducted extensive training at Fort Huachuca, Arizona—the only post in the United States with sufficient segregated facilities for the division's twenty-five thousand black men—prior to their deployment. Nicknamed the Buffalos and united under the motto Deeds Not Words, the men of the Ninety-second understood the importance of the division's performance to their race.

Inspection and awards ceremony for the Ninety-second Infantry Division at Viareggio, Italy, on October 20, 1944. (U.S. ARMY MILITARY HISTORY INSTITUTE)

Maj. Gen. Edward M. Almond assumed command of the Ninety-second and announced at its activation ceremony, "The Ninety-second Division is primarily a combat division. . . . One of my principal aims is to produce a first-class battlefield unit. I promise fairness to every officer and man; the best leadership of which your officers and non-commissioned officers are capable."

Almond, a decorated Regular Army veteran of World War I, certainly possessed the organizational and training skills to produce a combat-ready division. As time would tell, however, he may not have been the best man for the job. Almond, born in Virginia and graduated from the Virginia Military Institute, was an example of the War Department's unofficial stance that white southerners provided the best leadership for black units because they supposedly better understood the African-American individual and the race as a whole. What the War Department seemingly never realized was that these same southern white officers brought with them generational prejudice of long duration.

Neither Almond nor his predominately southern white officer corps ever developed confidence in their fellow black officers and enlisted men. Likewise, the black officers and enlisted men never developed a trust in Almond and his white officers. After a visit to the Ninety-second at Fort Huachuca in the summer of 1943, Brigadier General Davis noted: "General Almond had overlooked the human element in the training of the Ninety-second Division. Great stress has been placed upon the mechanical perfection in execution of training missions . . . and not enough consideration given to . . . maintenance of racial understanding between white and colored officers and men."

Not surprisingly, the Ninety-second did not initially perform in a stellar manner after landing on the Italian peninsula on July 30, 1944. Along the heavily defended Gothic Line in October, the experienced German troops stopped an attack by the Ninety-second Division and caused some of its battalions to retreat in a disorderly fashion. Although this was not unusual behavior for units in their first major engagement, especially when opposing veteran forces, once again rumors spread of black soldiers running from combat. By the time the story reached the press in the United States, periodicals such as *Newsweek* declared the "Luckless 92nd" a "disappointment and failure."

In March 1945, Truman Gibson, the civilian aide to the secretary of war, who had replaced Judge Hastie in January 1943 when Hastie resigned in disgust because of continued segregation and discrimination, inspected the Ninety-second Division and then met with reporters in Rome. Gibson confirmed to the press that the division had not done well in its initial fight but noted that it had performed bravely since, earning the praise of neighboring white units. He commented on the high rate of illiteracy in the division and problems in securing well-trained replacements because the War Department had made no plans for enlisting and preparing black reinforcements for combat units.

Gibson's remarks centered on his belief in the fighting ability and courage of black enlisted men and officers, but that is not what the white press focused on in their reports. In describing the poor performance of the Ninety-second in its first major fight, reporters quoted Gibson as referring to "panicky" retreats and the "melting away" of individual black soldiers from the front. Gibson would later deny ever making any statement about soldiers "melting away," but the damage had been done. Many whites in the United States readily believed the press reports, accepting a black official's remarks as proof of the inferiority of African Americans in combat.

Gun crew of Battery B, 598th Field Artillery, Ninety-second Infantry Division, preparing for a fire mission along the Arno River in Italy on September 1, 1944. (U.S. ARMY MILITARY HISTORY INSTITUTE)

For the rest of the war, the Ninety-second never "melted away"; its members earned more than twelve thousand individual decorations and citations and suffered more than three thousand casualties. However, their performance in World War II remained tainted by their first fight and Gibson's alleged remarks rather than celebrated for their subsequent six months of combat in which they fought as bravely as white units.

In addition to the three divisions, the army also formed several smaller black combat units in which African Americans proved that they could indeed learn to successfully operate sophisticated modern weapons. The 969th Field Artillery Battalion supported the defense of Bastogne and earned a Distinguished Unit Citation; the 452nd Antiaircraft Artillery Battalion supported the Allied drive across Europe into Germany; and the 614th Tank Destroyer Battalion stopped one of the German's last armor attacks near the town of Climbach, France, in 1944. Still another black unit, the 761st Tank Battalion, supported various white infantry divisions in their advance across France into Germany and Austria and fought in thirty major engagements during a period of 183 days of continuous combat. For its outstanding performance, the 761st received six nominations for the Presidential Unit Citation but did not actually receive the award until 1978 because of the prejudice and racism of senior white commanders.

Other black combat units trained for the war but never actually engaged the enemy. The 369th Antiaircraft Artillery Regiment provided defense in the Hawaiian Islands against the possibility of another Japanese aerial attack. In the United States black soldiers graduated from the elite Airborne School at Fort Benning, Georgia, and formed the 555th Infantry (Airborne) Battalion, with black officers filling all the command positions.

The "Triple Nickel," as the 555th became known, trained to join the Allied offensive in Europe. However, instead of deploying the unit to fight the Nazis, the army transferred them to the U.S. Pacific Northwest to fight forest fires in Washington and Oregon. Their only direct "war" action occurred when an occasional mission called for them to defuse firebombs, designed to start blazes in the American woods, which arrived aboard balloons launched by the Japanese into the easterly trade winds from their home islands.

Although they did not engage in direct combat, the morale and discipline of the 555th so impressed Maj. Gen. James Gavin, commander of the Eighty-second Airborne Division, that he invited the battalion to march with his unit in their victory parade down New York's Fifth Avenue on January 12, 1946, and then to join them at their new Fort Bragg, North Carolina, headquarters.

While black units served with distinction in the Pacific and Europe, they did so as segregated units. The nearest the U.S. Army came to integrating the fight occurred during the war's final months, once again as a result of the need for soldiers rather than any conscious move toward equality.

When the German offensive penetrated Allied lines in the Ardennes Forest, creating the Battle of the Bulge in December 1944, the United States suffered seventy-nine thousand casualties in a matter of weeks. The army rushed replacements from the States and converted whites in rear support units into infantrymen to fill the vacant ranks. When these measures still did not provide sufficient replacements, Lt. Gen. J. C. H. Lee, the chief of logistics for Supreme Commander Dwight

Eisenhower, recommended the integration of black support troops into the white infantry regiments. Eisenhower's chief of staff, Lt. Gen. Bedell Smith, spoke out against the move, declaring that it would violate official policy and encourage the black press and leaders in their movement to end racial segregation throughout the army. Lt. Gen. George Patton, commander of the Third Army, also entered the debate, expressing his belief that blacks did not possess the reflexes or intelligence to join his tank forces.

Eisenhower acknowledged the naysayers, but the need for manpower swayed his decision, and he issued a request "to all soldiers without regard to color or race" to volunteer for combat assignments. The order limited the number of black volunteers to twenty-five hundred but more than twice that number came forward. Some black sergeants willingly accepted a reduction in rank to become eligible to volunteer. Ultimately, thirty-seven black rifle platoons, led by white officers, joined the First and Seventh Armies. Patton refused to accept the black platoons, and none served in his command. In the First Army the black platoons joined white companies, creating the most integrated units of the war. The Seventh Army formed the black platoons into companies and then assigned them to white battalions to maintain a degree of segregation.

The black volunteer platoons fought side by side with the white units from March until Germany surrendered. In a postwar survey three out of four whites who served in units with the attached black platoons said that "their regard and respect for the Negro had risen." An official War Department study concluded that the black platoons "established themselves as fighting men no less courageous than their white comrades." Nevertheless, as soon as the war ended in Europe and the "need" for black replacements ceased, the army returned the platoons to their service-unit headquarters or discharged them.

Only black enlisted men had been allowed to volunteer for the platoons integrated into white units. Black officers remained in black units, where their numbers never came close to equaling the white officer-enlisted ratios. Throughout the war, young black men faced official and unofficial quota barriers to the officer candidate schools. By the end of the conflict only 7,768 black men had become officers. White officers represented more than 10 percent of the total number of whites in uniform, but less than 1 percent of blacks in the army received commissions.

All but a handful of these black officers received their commissions through the Reserve Officers Training Corps (ROTC) or in officer candidate school. Even though the United States was engaged in the largest war in its history and experienced its greatest demand for officers, the U.S. Military Academy graduated only a half-dozen black lieutenants during the entire conflict. Two of them graduated too late to participate in the war, but the other four all advanced in rank and earned decorations. One, Capt. Robert Tresville, of the class of 1943, died in aerial combat over Italy on June 22, 1944.

Once a black man did receive his commission, regulations limited him to service in black units, and both official and unofficial measures ensured that white officers would not serve under black superiors. Most African Americans remained lieutenants. Only Benjamin Davis Sr. joined the war's 776 generals, and a mere seven blacks were among the 5,220 men promoted to full colonel during the conflict.

While black enlisted men and officers suffered discrimination, black women, seeking to support their country's war efforts, faced inequality not only because of their gender but also due to the color of their skin. President Roosevelt signed Public Law 554 on May 14, 1942, authorizing the Woman's Army Auxiliary Corps (WAAC), and the War Department announced that they would accept female black officers and enlisted volunteers up to 10 percent of the corps' total. Despite these objectives, the WAAC's, converted to the Women's Army Corps (WAC) in July 1943, never had black women in excess of 5.9 percent of its members. At their numerical height during the last few months of the war, 120 black women officers and 3,961 black enlisted women were on active duty in the WAC.

Black WACs of all ranks, facing the same segregated facilities on army posts as black men, found themselves shut out of choice duty assignments and training schools and placed in menial positions regardless of their education or experience.

With the exception of one battalion, all black members of the WAC served in stateside assignments. Only after much protest by the black press and leaders did the single exception, the 6888th Central Postal Directory Battalion, sail for England in February 1945. Shortly before the war ended in Europe, the 6888th transferred to France, where they ensured the prompt delivery of mail to frontline soldiers and support troops.

Women's Army Corps officer candidates at Fort Des Moines, Iowa, in 1942.
(U.S. ARMY MILITARY HISTORY INSTITUTE)

Black women also volunteered for the Army Nurse Corps, but few found acceptance during the war's early years. By 1943 the nurse corps contained only 160 African-American women, all assigned to stations that had separate "Negro hospitals." Again, after much protest, the army declared that "every Negro nurse who puts in an application and meets the requirement" could join the corps. The National Association of Colored Graduate Nurses (NACGN) estimated that twenty-five hundred black nurses were available for enlistment, but the army, despite its promises, recruited only a few. By the end of the war, the Army Nurse Corps had only 479 black women—less than 1 percent of their total number.

A few black nurses did make it overseas, thirty of whom served in Monrovia, Liberia, in 1943 and treated black soldiers assigned there. A year later, sixty-three black nurses joined the 168th Station Hospital in England, where they cared for German prisoners of war. Still another group of fifteen black nurses served in Burma, Australia, New Guinea, and the Philippines during the final two years of the war.

Black women, enlisted men, officers, and volunteer infantry platoons and other African-American units received little recognition for their contributions. In fact, during all of World War II, black units and soldiers did not share in the decorations and honors heaped upon whites. Of the well over 1 million black men who served in the armed forces in World War II, not one received the Medal of Honor, the highest military award. Only nine received the second-highest honor, the Distinguished Service Cross.

For nearly a half century the only explanation provided by the U.S. military for this oversight of African-American individual bravery was that black units experienced limited combat in the war. Black leaders blamed the whites' belief in the inferiority of African Americans for the discrepancy. The real reason why no black man received the country's highest award was on the front cover of the May 6, 1996, edition of *U.S. News & World Report* in a bold headline: "Military Injustice: No black soldier received the Medal of Honor, America's highest award for valor, during World War II. The reason was racism."

The catalyst for the magazine story was the Department of Defense recommendation that seven black soldiers be awarded the Medal of Honor. The three-year Department of Defense study that made the medal recommendations also provided perspectives into race relations during World War II: "Segregated units by race complicated and slowed training, exacerbated relations between officers and enlisted men and between commanders and their units, and undermined the morale of these units in both subtle and obvious ways."

Included in the study were remarks by Lt. Gen. (Ret.) William McCaffrey, who served as a white officer in the ranks of captain through colonel under Gen. Edward Almond in the Ninety-second Infantry Division during World War II. McCaffrey said, "Almond came out of Luray, Virginia, and he had the attitudes of that time and that place. Hell, everyone in the army then was a racist."

Racism aside, the valor of these seven enlisted and commissioned African-Americans certainly merited such recognition. Pvt. George Watson, of the 29th Quartermaster Regiment, rescued several fellow soldiers from drowning when Japanese bombers sunk their transport near New Guinea on March 8, 1943. Watson sacrificed his own life when he crawled back aboard to rescue more soldiers and succumbed in the turbulence of the sinking ship.

On November 19, 1944, Staff Sgt. Ruben Rivers, of Company A, 761st Tank Battalion, although seriously wounded, refused evacuation during a battle near Guébling, France. Rivers remained in his tank and stopped a German counterattack, protecting the remainder of his company. Before the German tanks overwhelmed and killed him, comrades heard Rivers shout, "I see 'em. We'll fight them."

Black officers also displayed their leadership and valor. On April 5, 1945, 1st Lt. Vernon J. Baker led his twenty-five-man African-American platoon of Company C, 370th Infantry Regiment, Ninety-second Infantry Division, against heavily fortified German positions at Castle Aghinolfi, Italy. Baker charged an observation post, killing its two German defenders with his rifle. Continuing the assault, Baker and his platoon destroyed six enemy machine-gun positions and killed twenty-six German soldiers. Only Baker and six of his men survived the intense battle.

Uncommon valor was common in other African-American units. First Lt. John R. Fox, a forward observer in the 366th Infantry Regiment, Ninety-second Infantry Division, called for artillery fire on his own position on December 26, 1944, in northern Italy and destroyed a large German force threatening the remainder of his regiment. Near Speyer, Germany, on March 23, 1945, Staff Sgt. Edward A. Carter Jr., of the Fifty-sixth Armored Infantry, Twelfth Armored Division, killed six German soldiers and captured two more in what became hand-to-hand combat.

On April 7, 1945, PFC Willy F. James Jr. of Company G, 413th Infantry Regiment, 104th Infantry Division was killed while attempting to rescue his wounded platoon leader from German fire at a bridgehead across Germany's Weser River. First Lieutenant Charles L. Thomas, Company C, 614th Tank Destroyer Battalion, 103rd Infantry Division, although seriously wounded near Climbach, France on December 14, 1944 refused evacuation and directed the defense that saved the remainder of his company.

More than fifty years after their brave deeds, the seven black soldiers finally received the recognition they so richly deserved. Based on the Department of Defense study, and encouraged by media exposure, the U.S. Congress in October 1996 authorized the medals as a part of the 1997 defense appropriations bill.

In a White House ceremony on January 13, 1997, President Bill Clinton presented the Medal of Honor to the only survivor, Vernon Baker, and to the families of the other six heroes. Standing before the famous Gilbert Stuart portrait of George Washington, President Clinton declared, "History has been made whole again, and our nation is bestowing honor on those who have long deserved it. They were denied their nation's highest honor, but their deeds could not be denied, and they cleared the way to a better world."

Arlene Fox, widow of John Fox, remarked at the ceremony, "It's a long time coming," and added she held no ill will against the army, concluding, "I don't dwell in negativism. It's a very proud day."

When asked about risking his life for his country while serving in a segregated unit, Baker replied, "I was an angry young man. We were all angry. But we had a job to do, and we did it."

11

WORLD WAR II: ARMY AIR FORCE

Between the two world wars few things inspired the imagination of the American public as much as aviation. Charles Lindbergh became an overnight celebrity with his pioneer solo crossing of the Atlantic Ocean in 1927. Then former World War I pilots and newly qualified aviators brought flight closer to home by "barnstorming" across the United States using farm fields and country roads as runways. They put on flying demonstrations and provided local citizens their first airplane rides. Soon motion pictures, comic strips, and children's books featured daring pilots as their heroes.

African Americans avidly participated in aviation from its very beginnings, although whites rarely acknowledged their accomplishments. Dale White and Chauncey Spencer flew from Chicago to Washington, D.C., in May 1939 to demonstrate to white nonbelievers that blacks could indeed fly. In 1932, Charles A. Anderson became the first black pilot to make a round-trip transcontinental flight, and in 1939 he opened his own commercial air service in Washington, D.C.

Black women also climbed into the cockpits, and during the early 1930s the Bessie Coleman Aerocircus put on shows throughout the West and Southwest; Willa Brown operated the Coffey School of Aeronautics in Chicago. By 1939, 125 African Americans held pilot licenses.

Black pilots, again mostly unknown to white America, had also proved their abilities in combat prior to World War II. In addition to Eugene J. Bullard, who had flown with the French in World War I,

J. C. Robinson, after a barnstorming tour in the United States, sailed for Ethiopia, where he flew for that country against the Fascist Italian invaders in the late 1930s. In less than four months of flying as a volunteer in the Spanish Civil War, another African American, James Peck, became an "ace," with five aerial-combat victories.

Despite civilian and combat demonstrations of competence, African Americans found military flying closed to them. The air corps, which considered itself the elite of the modern American military, remained restricted to whites only. Despite evidence to the contrary, many white Americans, including high-ranking government officials, refused to believe that blacks could pilot sophisticated aircraft.

Secretary of War Henry Stimson admitted that he thought black pilots could make no contribution to the war effort and considered their potential only as a political concession rather than as a viable tactical option. Stimson's assistant secretary of air, Texas-born Robert A. Lovett, also held racist beliefs about black aviators. When informed that there was no scientific evidence that blacks could not fly as well as whites, Lovett responded, "No, I don't suppose so; there must be some emotional reason."

With such strong bias entrenched in white senior American military decision makers, it is not surprising that blacks in World War II had to fight for the right to fly just as they had had to fight for the right to fight. Only through organized protests by black leaders and newspapers did African Americans gain an open door to the air corps.

The process, however, was not easy; nor did it ever reach any great degree of equality. Public Law 18, enacted on April 3, 1939 called for contracts between the army air corps and civilian colleges and universities to provide flight training for military pilots. To appease black leaders and Eleanor Roosevelt, the law included provisions for at least one black school to train pilots.

The air corps, objecting to college-trained black pilots seeking enlistment, moved to block their entry. On April 8 the chief of the air corps plans division announced that the law only required the army to lend aircraft and equipment to the schools for training of black pilots but that it did not require the air corps to accept them into uniform once they became qualified aviators.

On June 27, 1939, Congress passed the Civilian Pilot Training Act (CPTA), which provided more opportunity for blacks to earn their pilot wings. Intended to create a large reserve of qualified civilian aviators in case the United States entered the war in Europe, the act opened pilot training to black students at Delaware State, Howard University, Hampton Institute, North Carolina A&T, the University of Missouri, West Virginia State College, and Tuskegee Institute. Of the first one hundred black students to enroll in the CPTA, ninety-one successfully completed the course and earned their private pilot licenses.

Hundreds of African Americans wanted to become military pilots, and Public Law 18 authorized their training. But the air corps stood by their policy of excluding blacks. Even after the War Department authorized the North Suburban Flying School in Glenview, Illinois, which worked closely with Willa Brown's aviation-instruction organization in Chicago, to train black military pilots, the air corps continued their resistance.

The air corps maneuvered again on May 25, 1940, to block blacks, this time using the issue of integration:

> We are having difficulty in finding twenty qualified students needed to begin instruction, and now that the War Department is funding and financing the program, and it is the policy not to mix colored and white men in the same tactical organization, and since no provision has been made for any colored air corps units in the army, colored persons are not eligible for enlistment in the air corps, so there is no need for additional facilities.

This convoluted way of saying that the U.S. Air Corps did not want African Americans offended the black community and their white supporters, including Mrs. Roosevelt. Nevertheless, the air corps did not relent in its policy of excluding blacks until forced to do so by the Selective Training and Service Act of September 14, 1940, which prohibited discrimination based on race or color in the selection, induction, and training of military personnel.

Two days after the act passed, the air corps announced that it would accept blacks in compliance with the new law. When forced to do so, the air corps already had a plan to maintain segregation and to keep black pilots out of the cockpit, avoiding the spirit, if not the letter, of the legislation. Four months earlier, on May 31, Gen. Henry H. ("Hap") Arnold, chief of the air corps, declared in a memo that if required to accept blacks, the air service would use them "in labor battalions or labor companies to perform the duties of post fatigue and as waiters in our messes."

One of the few Americans ever promoted to the rank of five-star general, Arnold also explained in his memo that only enlisted positions would be open;"Negro pilots cannot be used in our present air corps units since this would result in having Negro officers serving over white enlisted men. This would create an impossible social problem."

African Americans adamantly protested their exclusion from the glamorous and prestigious air arm and continued their efforts to integrate the other branches of the armed services as well. Black pressure, supported by a lawsuit filed by Howard University student Yancey Williams with the help of the NAACP, finally forced the War Department on January 9, 1941, to authorize the training of black pilots and form the Ninety-ninth Pursuit Squadron.

According to the plan, the Ninety-ninth Pursuit Squadron and all other black air corps support units would operate as "separate" units outside the white organizations. The Tuskegee, Alabama, airfield, with a CPTA program already established, would serve as the headquarters for black-aviator training.

The plan satisfied no one. Many in the War Department and the air corps remained skeptical that blacks could learn to fly. The African-American community welcomed the opportunity for black men to become military pilots, but they resented the fact that black aviation units, like the rest of the army, would be segregated.

From the beginning of the "Tuskegee Experiment," both whites and blacks realized that the success or failure of the program's graduates would determine the future of African-American aviators in the army. On August 25, 1941, the Tuskegee School received the first class of thirteen black cadets. For the next six months the men trained stringently both in the classroom and in the air, for blacks and whites agreed that only the very best should graduate and become leaders of the black squadron.

Five of the cadets completed the training in March 1942 and pinned on their air corps pilot wings. Among the graduates was Benjamin O. Davis Jr., member of the West Point class of 1936 and son of the army's only black general. The younger Davis had earlier sought pilot training after graduation from the U.S. Military Academy, only to be turned down and assigned to the Twenty-fourth Infantry Regiment. When

The first graduating class of Tuskegee Airmen (Benjamin O. Davis Jr., *third from left*), March 7, 1942. (U.S. ARMY MILITARY HISTORY INSTITUTE)

the Tuskegee School began accepting applications, he had been one of the first to apply. Because of his five years of previous service as an infantry officer and his performance in flight training, Davis took command of the Ninety-ninth Pursuit Squadron after graduation, for no senior black airman existed.

Subsequent classes produced additional black pilots for the Ninety-ninth Pursuit Squadron. In addition to pilots, black air corps support personnel trained at Tuskegee as well as at other installations. By mid-1941 a total of 2,250 had completed training; a year later the number reached 77,500; and by the end of the war it was 145,000.

By any measurement, proportional or otherwise, the six hundred African Americans who earned their pilot wings at Tuskegee were little more than tokens. The Tuskegee airmen, however, proved they could fly, and now they wanted to show that they could fight.

By October 1942 enough pilots and ground crews had completed training to fully man the Ninety-ninth Pursuit Squadron. A white Third Air Force inspector reported that the squadron was "in excellent condition and ready for immediate overseas shipment." Although the squadron was ready to fight, the War Department was not prepared to send the black fliers into battle. Reluctant to train them in the first place, the War Department had never made any plans to actually use the black squadron in combat. The nearest they had come to committing the black squadron overseas had been a plan to deploy them to Liberia, a country originally formed by freed American slaves in a "back to Africa" movement. The plan called for black aviators to fly offshore missions against German submarines preying on Allied shipping lanes.

Once more black leaders and supporters protested. This was the issue that led Judge Hastie to resign as assistant secretary in January 1943. On March 30, 1943, Eleanor Roosevelt wrote to the air staff to express her surprise about the nondeployment of the black squadron and ask, "Does this mean that none of those trained are being used in active service?"

These factors and the age-old American tradition of allowing blacks to enter combat only in the face of dire need paved the way for the Ninety-ninth to join the fight. The Ninety-ninth finally arrived in French Morocco on April 24, 1943—black American pilots were now on the African continent to fight for the United States against the Axis powers.

For a month the now-redesignated Ninety-ninth Fighter Squadron trained on their own, with little assistance from white squadrons of the Thirty-third Fighter Group at their base outside the town of Fez. The African-American pilots found that few white aviators took them seriously. Instead of receiving any personal or professional courtesy from the whites, black fliers endured such epithets as "boy" and "nigger."

With no veteran pilots in their squadron, the officers of the Ninety-ninth had much to learn. Black pilots had particular difficulty in cross-country navigation because of the seemingly endless, bare desert landscape and their own inexperience. While white pilots had learned cross-country navigation before being deployed from the United States by actually flying from base to base, the Tuskegee airmen had missed this valuable training because most stateside installations, with no separate facilities for blacks, refused to let them land and use "whites only" quarters and messes.

Nevertheless, the men of the Ninety-ninth learned quickly and flew their first combat mission over the Mediterranean island of Pantelleria on June 2, 1943. A few weeks later, the Ninety-ninth transferred to the 324th Fighter Group, based at El Haouaria, Tunisia. On July 2, the squadron escorted a formation of B-25 Mitchell bombers to the Sicilian coast, where they came under attack by German FW-190s. In the ensuing dog fight, Lt. Charles Hall of Brazil, Indiana, scored the Ninety-ninth Squadron's first kill of an enemy aircraft.

Gen. Dwight D. Eisenhower, then commander of American forces in North Africa, and high-ranking air corps officers met Hall and his comrades upon their return from the mission to congratulate them. Announcement of the first aerial victory by a black pilot also received notice in the American press. The July 18, 1943, issue of the *Birmingham News*, stated: "When the screaming P-40 Warhawks, piloted by the first Negro fighter pilots in the history of the world, roared through the Mediterranean skies to aid an allied offensive . . . , the Tuskegee trained pilots faced their acid test and came through with flying colors to prove that they had the necessary mettle to fly successfully in combat."

The Ninety-ninth Squadron was off to a good start, but senior white air corps officers, who objected to integration, thought they still had much to prove, leaving the black pilots to confront German adversaries in the skies and American racism on the ground. White commanders in North Africa noted in their reports that the Ninety-ninth lacked aggressiveness. Maj. Gen. Edwin House, commander of the XII Air Support Command, wrote in a September 16, 1943, report that "the Negro type has not the proper reflexes to make a first-class fighter pilot."

Army air force officials in Washington, including Gen. "Hap" Arnold, welcomed the negative reports about the black pilots despite their vagueness and nonspecificity. In September, Arnold sent a memo by special courier to Lt. Gen. Carl Spaatz, the senior air commander in North Africa, urging that more adverse reports about the Ninety-ninth be submitted to justify not increasing the number of black pilots in the air corps. Arnold wrote, "We have received from very unofficial sources second-hand tales of the fact that the Negro tires easily and that he loses his will to fight after five or six missions. . . . I am sure that you realize the urgency required for the information, in view of the fact that we contemplate building additional Negro units at once."

Arnold and the air force were well on their way to disbanding the Ninety-ninth, or at least assigning it to a noncombat area, when the

War Department placed the matter before its Advisory Committee on Negro Troop Policy, headed by Assistant Secretary John McCloy. The committee, consisting of representatives of the army's ground, air, and service forces, heard the complaints from the air corps staff and then comments from Benjamin Davis Jr., the former commander of the Ninety-ninth, now back in the States to organize a new black aviation unit.

Davis met the controversy head-on, outlining the Ninety-ninth's limited training and its lack of veteran pilots upon arrival in North Africa. He then declared that the accusations of the squadron's pilots "tiring easily" were untrue. Davis pointed out that their fatigue came from simply flying more missions than the white squadrons because the Ninety-ninth had only twenty-six pilots compared to thirty-five in the white squadrons to fly the same number of aircraft. As for aggressiveness, Davis noted that despite the diminishing number of enemy aircraft in the Mediterranean theater, the Ninety-ninth had engaged the enemy on 80 percent of its missions.

Based on Davis's testimony, the committee recommended the retention of the Ninety-ninth and the increase in black squadrons. Still, Arnold continued plans to eliminate, or at least diminish, black air units. Not until members of his own staff pointed out the potential political ramifications did Arnold reconsider. Politics, along with reports of additional successes by the Ninety-ninth Squadron, finally convinced Arnold to cease efforts to end the "Tuskegee Experiment."

By January 1944 the now-veteran black pilots of the Ninety-ninth Squadron supported the Anzio invasion and scored eight confirmed kills of German aircraft—the highest number by any squadron in the operation. Even the air corps now had to admit that black pilots could indeed fly. *Time* magazine also recognized the prowess of the Ninety-ninth at Anzio in their February 14, 1944, issue: "Any outfit would have been proud of the record. These victories stamped the final seal of combat excellence on one of the most controversial outfits in the army, the all-Negro fighter squadron."

In defiance of the air corps' rejection of them, black pilots of the Ninety-ninth adopted the unofficial name "Spookwaffe"—a combined form of the German for air force, "Luftwaffe," and the racist term "spook," often used in place of "nigger" by whites. Pilots of the Ninety-ninth enjoyed the displeasure senior white officers displayed toward a nickname that also symbolized their "invisible" status. Both in the air and on the ground, members of the Ninety-ninth Squadron wore yellow scarves boldly embroidered with their "Spookwaffe" nickname.

While the Ninety-ninth added to their victories in the skies over Europe, Colonel Davis assumed command of the 332nd Fighter Group at Selfridge Field, Michigan, in October 1943. The all-black 332nd, activated at Tuskegee Army Air Field on May 26, 1942, contained the 100th, 301st, and 302nd Fighter Squadrons.

Davis, now a veteran combat pilot, rigorously trained his squadrons before their deployment to Italy in February 1944. On March 17 the 332nd engaged in its first aerial action against the enemy. In July the Ninety-ninth Fighter Squadron joined the 332nd, making this unit the only four-squadron fighter group in the air force and, of course, the only unit with black pilots. From the time of the assignment of the Ninety-ninth to the group until the end of the war, the 332nd flew bomber-escort duty, including missions deep into the German heartland.

The 332nd had the tail section of each plane painted bright red, prompting the bomber crews they escorted to call them the "Red Tails." These white bomber crews were more than happy to look out over their wing tips and spot the distinctive red markings, because the 332nd reputedly was the only escort group never to lose a bomber to enemy fighters during the war.

The black pilots of the Ninety-ninth Squadron and their comrades in the 332nd Fighter Group earned both individual and collective honors. Flying P-39, P-47, and P-51 fighters, the Tuskegee airmen flew a total of 1,578 missions, consisting of 15,533 combat sorties, during which they destroyed 261 enemy aircraft in the air or on the ground and damaged another 148. Black pilots earned ninety-five Distinguished Flying Crosses. The Ninety-ninth Fighter Squadron earned three Distinguished Unit Citations; the 332nd Fighter Group, one. The group lost sixty-six men in action.

At the conclusion of the war in Europe, Colonel Davis and a cadre of forty officers and enlisted men of the 332nd returned to the United States to assume command of the 477th Composite Group and prepared to join the fight against Japan. The 477th, originally organized as a bombardment group in June 1943, had never received the manpower or the training priorities to become combat ready. During its first year, the group's squadrons moved thirty-eight times and always lacked enough assigned pilots, navigators, and bombardiers to conduct proper training.

This turbulence and the low priorities destroyed the spirit of the men of the 477th. The discrimination they experienced both on and off base also further eroded their morale. In April 1945, at Freeman Field, Indiana, the air corps arrested 101 black officers of the group for refusing to comply with a regulation restricting their access to an officers' club. Roy Wilkins, writing in the May issue of the *Crisis*, best summarized the incident: "The 101 young Negro officers arrested at Freeman Field . . . are determined to be treated as officers, not as Negroes. If there is a club for officers, then they insist to use it. If attempting to use it means arrest, then they choose arrest."

The black officers remained in confinement until April 26, when the War Department directed the air corps to dismiss charges. Morale remained so low in the 477th that even Davis and his veteran staff could not prepare the group for combat before Japan surrendered to end the war.

While only the 332nd Group actually faced the declared enemy during World War II, they, along with the 477th and other black men and women in uniform, also fought racism and discrimination in their efforts to defend their country. Participation by the Tuskegee airmen, limited by the air corps and not by their desire to serve, had proved that blacks could indeed fly and fly well. No longer would the U.S. armed forces be able to maintain segregated skies.

12

WORLD WAR II: NAVY AND MARINE CORPS

On the morning of December 7, 1941, Doris "Dorie" Miller, a black messman, awoke early to collect dirty laundry aboard the battleship USS *West Virginia*, which lay at anchor in Hawaii's Pearl Harbor. He was going through his regular routine when he suddenly felt and then heard terrifying explosions. Quickly making his way topside, Miller discovered mayhem when he stepped on deck. The ship's bridge was on fire, and bombs and aerial torpedoes were raining from the skies. Through the chaos of screams, shouts, and reverberating blasts Dorie Miller raced to the burning bridge, where he discovered his wounded captain. Miller dragged the officer from the inferno to safety and then turned to survey the situation.

By the time the next wave of enemy aircraft swooped from the skies to continue the attack, Miller was manning an antiaircraft gun. Despite having no training with the weapon, it was confirmed that he shot down two Japanese planes and possibly two more.

Dorie Miller's heroism went unheralded for nearly a year. The navy then bestowed on this son of a Texas sharecropper the Navy Cross, the service's second-highest valor award, for "distinguished devotion to duty, extraordinary courage, and disregard for his own personal safety." Miller also received a promotion from mess attendant second-class to mess attendant first-class. Two years later, Miller, still waiting on officers in the mess and gathering their dirty laundry aboard the light aircraft carrier USS *Liscome Bay*, went down with his ship in the South Pacific when a Japanese submarine torpedoed it on November 24, 1943.

At the time of Pearl Harbor, the five thousand black sailors on active duty were, like Miller, almost exclusively restricted to working as mess attendants. Although pressure from black leaders, organizations, and newspapers had forced the ever politically astute Roosevelt to encourage the navy to open its ranks to black enlistees, the staunch senior commanders resisted.

In an effort to keep blacks in subservient positions, Secretary of Navy Frank Knox expressed doubts that African Americans had the ability to assume regular sailor duties. Knox even quoted the results of a failed post–World War I experiment to man ships with Filipino and Samoan crews as an example of the inability of people of color to become efficient sailors.

Knox took action only after Roosevelt insisted. The secretary directed the General Board of the Navy, composed of senior naval and marine officers, to devise a plan to assimilate an additional five thousand black recruits into the navy in positions other than as messmen. The board members studied the issue of more African Americans joining their commands only with great reluctance. Their attitude was blatant. In the January 8, 1942, meeting of the board, Gen. Thomas Holcomb, commandant of the Marine Corps, noted, "The Negro race has every opportunity now to satisfy its aspirations for combat in the army—a very much larger organization than the navy or marine corps—and their desire to enter the naval service is largely, I think, to break into a club that doesn't want them."

The other members of the General Board shared Holcomb's desire to keep their club as white as possible, and on February 3 the group issued a memorandum stating that they were unable to comply with the secretary's directive to accept black enlistees in any specialty other than the messman branch. According to the board, segregation was a fact of life not only in the military but also throughout the United States and its possessions. The board reported that integration of the navy simply would not work, explaining:

> (a) The white man will not accept the Negro in a position of authority over him; (b) the white man considers that he is of a superior race and will not admit the Negro as an equal; and (c) the white

> man refuses to admit the Negro to intimate family relationships leading to marriage. These concepts may not be truly democratic, but it is doubtful if the most ardent lovers of democracy will dispute them, particularly in regard to inter-marriage.

While reasons (a) and (b) accurately reflected the general feelings of whites in the navy, the board apparently added (c) in an attempt to inflame the racism of their superiors. The board's only charter was to study how blacks could achieve equality in the navy. The question of intermarriage was obviously beyond their agenda and had nothing to do with naval proficiency.

The board members concluded that the current policy should continue and referred the final decision back to the president: "If in the opinion of higher authority, political pressure is such as to require the enlistment of these people for general service, let it be for that."

"Let it be for that" was an ambiguous phrase subject to interpretation. Some thought it meant that the board was suggesting opening the general service to blacks, an option beyond even Roosevelt's support of equality. Others interpreted the words to mean that the acceptance of blacks was a political rather than a military issue.

As the "higher authority," Roosevelt made clear his opinion to Secretary Knox. In a memo to the General Board on February 14, 1942, Knox expressed Roosevelt's reaction: "The President is not satisfied with the alternative suggested by the recent decision of the General Board. He thinks that some special assignments can be worked out for Negro enlisted men which would not inject into the whole personnel of the navy the race question."

The General Board went back to work and under Roosevelt's scrutiny finally announced, on April 7, 1942, that blacks could enlist in the navy for service other than as mess attendants. In fact, thanks to the board, the navy suddenly had a wide range of specialties open to blacks. On June 1 black volunteers began training in segregated classes at bases in Illinois and Virginia as gunners, clerks, radio operators, and signalmen. Few, however, actually used their training after graduating, since the navy did not assign them to positions where they could practice their newly acquired skills. Instead, the navy put its newly trained specialists in ammunition facilities or construction battalions. Black leaders complained bitterly about the hoax, saying that the only advancement had been in black sailors trading their waiters' aprons for the carpenter's hammers and stevedore's hooks.

During the first six months after the "opening" of specialties, few blacks volunteered for naval service, the vast majority joining the army instead. Naval authorities, not unhappy with the small number of black enlistments, justified the trend as the result of blacks' natural "fear of water" and their lack of knowledge about the sea. Black leaders countered that young African Americans viewed the navy as even more racist than the army and knew about the limited opportunities aboard ships. Whatever the reason, black volunteers avoided the navy to such an extent that by February 1943 only twenty-six thousand African Americans wore navy uniforms. More than two-thirds of the black sailors, roughly eighteen thousand, still held the rating of mess attendant.

The numbers of blacks in the navy remained at about 2 percent of the total force until that pivotal point of February 1943, when the War Manpower Commission announced that all the services had to accept a fair percentage of draftees and end the navy's policy of accepting only volunteers. The navy protested the policy, explaining it could only accept a small percentage of the black draftees. President Roosevelt once again exercised his higher authority and directed Knox to comply with the directive and enlist a full share.

By the end of 1943 black naval enlistees numbered 78,000 of the total of 740,000 accepted into the service. Of the 78,000 inducted in 1943, fully 38,000 of the black sailors were assigned as mess attendants. Even though more than 10 percent of the new inductees were black, the influx was not enough to offset the previous years' low numbers in terms of cumulative ratios of blacks to whites. Despite the directives of President Roosevelt and the largest military manpower requirements in U.S. history, only 150,000 African Americans, or about 5 percent of the total, served in the U.S. Navy during World War II.

Roosevelt had directed the navy to comply with the War Manpower Commission, but he had not specifically demanded the navy to integrate its crews to accommodate more blacks. Accordingly, the navy had agreed to accept the increase, but senior officers were not, of their own volition, interested in initiating integration. To avoid the issue, the navy continued its practice of training blacks in the expanded specialties and then assigning them to manual-labor jobs, like hauling ammunition or building roads and docks, rather than actually utilizing the skills they learned in training.

Not surprisingly, the navy's actions garnered complaints from black leaders and created racial tension on naval installations, some of which led to physical confrontations. With a need to "calm the waters," navy officials devised a plan to present a show of good faith. They established a Special Programs Unit in August 1943 to oversee the enlistment and assignment of African-American sailors.

To oversee the unit, the navy chose four relatively junior officers, with the knowledge that these ranks had little military or political clout. The navy could then point to its special unit as a sincere effort at improving the quality of life for black sailors, all the while assured that the unit had no real power. The navy erred, however, in its choice of officers for the Special Programs Unit.

The four junior white officers, the principal staff of the Special Programs Unit, included Capt. Thomas F. Darden, head of the organization, and Lt. Comdrs. Donald O. VanNess, Charles E. Dillon, and Christopher S. Sargent, his subordinates. VanNess and Dillon brought to the unit their experience as having established and run the first general specialty schools open to black enlistees. Sargent, an energetic advocate of racial equality, also maintained his prewar ties with influential business and political leaders to exert pressure on his superiors to provide additional opportunities for African-American sailors.

Instead of merely going through the expected, perfunctory motions of responding to complaints, the Special Programs Unit began to study the overall racial situation in the navy and to make recommendations for the increased use of African Americans. The unit worked to devise methods to use black recruits in the most efficient manner as it steered the navy toward a course of increased equality for its minority sailors. Even though the unit failed during its first year of operation to secure general integration, it did achieve a token accomplishment in gaining approval of black sailors manning several ships.

Interestingly, with all the memos and stacks of reports written during World War II concerning the acceptance of additional blacks into the navy, little documentation exists about the approval of the experiment to man an escort destroyer and a submarine chaser with black crews. The first mention in the archives about the matter is a memo from the Office of Naval Personnel to the commander of the U.S. Fleet dated December 1, 1943, announcing the manning of the two vessels. The next was the public announcement on February 23, 1944, that the ships were about to be launched with their new crews.

Secretary Knox may have verbally approved the decision. No records indicate whether he conferred with the president. Perhaps in the chaos of war no one recorded these conversations or made note of the decision, or maybe such documentation has since been lost or misfiled. Another explanation for the lack of records about the decision to man two ships with black crews could be that no one wanted to shoulder the responsibility if the experiment failed.

The escort destroyer destined for a black crew was the USS *Mason* (DE-529), built at the Bethlehem Shipyard in Quincy, Massachusetts, and commissioned on March 20, 1944. At 1,140 tons and 290 feet in length, the *Mason*, like other destroyer escorts, had the single mission to protect convoys from enemy submarines and to take a torpedo themselves rather than allow the larger cargo ships to be hit. The *Mason*'s crew of 204 consisted of six white officers, 38 white petty officers, and 160 black crewmen. Except for command, black sailors served in every shipboard capacity.

A little more than a month after the commissioning of the *Mason*, the USS *PC-1268* was launched from the Consolidated Shipbuilding Company yard in Morris Heights, New York, on April 24. The other black-manned navy ship, 173 feet in length and with a displacement of five hundred tons, may have been smaller in size than the *Mason*, but its accomplishments were just as impressive.

In July 1944 the subchaser *1264* began escort duty for convoys along the East Coast of the United States, from the northeastern states to Florida and Guantanamo Bay in Cuba. The original crew of the *1264* consisted of five white officers and fifty-eight enlisted men—fifty blacks and eight white petty officers. According to the plans of the Special Programs Unit, which monitored the progress of the *1264* as well as the *Mason*, the white petty officers were to remain onboard only long enough to train the inexperienced black crewmen. Late in 1944 the unit and Lt. Eric S. Purdon, the captain of the *1264*, agreed that the black sailors were ready to assume all crew duties. African-American seamen aboard the *1264* were promoted to petty officers and replaced white noncommissioned offices,

Crew members from the USS *Mason* (*background*) at Boston Navy Yard on March 29, 1944. (U.S. NAVAL INSTITUTE)

who transferred to other ships or shore duties. For the remainder of the war the enlisted crew of the *1264* remained all black.

During its wartime service the *Mason* made six voyages across the Atlantic, escorting supply convoys. In October 1944 the *Mason*, as a part of Convoy NY-119, consisting of ships and barges sailing from New York to Portsmouth, England, exceeded all expectations for any crew of any race.

As the convoy neared the English coast, it encountered a severe storm that scattered the ships and broke loose several of the towed barges. The *Mason* and two British destroyers managed to safely lead about half of the convoy into a protected harbor at Falmouth. Pausing only to conduct emergency welding repairs to deck seams separated by the storm's massive waves, Capt. William Blackford turned the *Mason* back toward the open sea and joined two British destroyers in securing the other cargo ships and barges. The convoy's senior officer, Comdr. Alfred Lind, advised his subordinate captains to proceed at their own discretion. When the three escorts encountered forty-foot waves as they sailed from the harbor channel into the open sea, the British vessels turned back. The *Mason* continued, and despite sustaining additional damage, located the remainder of the convoy and led it to safe moorage in Falmouth harbor.

In his official report, Commander Lind praised the performance of the *Mason* and its crew and recommended "that this ship be given a letter of commendation to be filed in the record of each officer and man on board that vessel."

Captain Blackford also recommended his crew for a commendation, but the navy took no action to recognize the *Mason* and its black crew. Even the navy's final report at the end of the war downplayed the proficiency of black sailors, declaring only, "The USS *Mason* served in Atlantic convoy duty [and] operated satisfactorily."

While the black-manned *Mason* was escorting convoys across the Atlantic, the *1264* and its all-black crew focused on its coastal mission of submarine surveillance. During the last six months of the war against Germany, the *1264* performed independent antisubmarine patrols off the East Coast and was preparing to redeploy to the Pacific when Japan surrendered.

On October 27, 1945, the fleet celebrated Navy Day with a huge demonstration of naval might in New York harbor, with President Harry Truman as the reviewing officer. The navy selected forty-seven ships, including aircraft carriers and battleships, for the demonstration. The smallest in size but not in the hearts of its crewmen was the *1264*.

Veterans of the *Mason* and the *1264* engaged in a friendly rivalry during and after the war over the proficiency of their ships and crews. The *Mason* crew hailed their six transatlantic passages, while the *1264* veterans proudly noted that they alone had earned the right to replace the white petty officers with men of their own race. Both crews had more than enough reason to be proud, for their performances had been exemplary. The navy, however, took little note of the successful "experiment" and, in the postwar reduction of men and ships, quickly and quietly decommissioned the *Mason* and the *1264*.

The crews of the *Mason* and the *1264* proved that black sailors could perform every shipboard duty well. However, their officers remained all white for the simple reason that the navy did not commission its first black officer until 1944.

To meet the leadership requirements of the rapidly expanding fleet, the navy initiated, on July 1, 1943, an expedited program, known as V-12, to combine college education and officers training. Black leaders pressed for V-12 to accept African-American officer candidates, but the Bureau of Naval Personnel and Secretary Knox resisted. President Roosevelt finally intervened and directed that black students be included in V-12.

Assistant Secretary of the Navy Adlai Stevenson, the future governor of Illinois and two-time candidate for president, proposed a more immediate solution in a memo to Secretary Knox on September 29, 1943. Stevenson recommended an accelerated commissioning program for "10 or 12 Negroes selected from the top-notch civilians just as we procure white officers, and a few from the ranks." In his memo Stevenson justified his recommendation, declaring, "I feel very emphatically that we should commission a few Negroes. . . . Ultimately there will be Negro officers in the navy. It seems to me wise to do something about it now. . . . I don't believe we can or should postpone commissioning some Negroes much longer."

Secretary Knox concurred with Stevenson's recommendation, and over the next few months the navy carefully screened black recruits as well as those already on active duty. On January 1, 1944, the sixteen selected candidates began their classes in segregated facilities at Great Lakes Naval Training Station, Illinois.

Like the authorization of the two black-manned ships, little documentation has surfaced in the archives about the selection of the black sailors or their training. From the existing evidence, it appears that the navy conducted the accelerated training program in good faith and encountered few difficulties.

The navy had accepted sixteen black men into the officers training program, anticipating a 25 percent attrition rate. At the end of the ten-week course, all sixteen had satisfactorily passed all aspects of the training. Even so, the navy decided to commission only twelve, as originally planned. Upon graduation, those twelve black men received commissions as ensigns in the U.S. Navy Reserve. As a consolation, one candidate, who lacked a college degree, received an appointment as a warrant officer. The other three retained their enlisted rank and rejoined their units.

Little fanfare accompanied the commissioning, the navy wishing to downplay the breach in its all-white officers' club. It was not until the

Twelve of the "Golden Thirteen": The first black officers commissioned in the U.S. Navy, 1944. (U.S. Naval Institute)

graduates held a reunion in 1977 that they dubbed themselves the "Golden Thirteen"—the name by which the navy, twenty-five years after the fact, would promote its World War II racial-equality program.

All thirteen received stateside assignments, six remaining at Great Lakes to train black recruits and the others assuming patrol or tugboat duties in the New York, Boston, and San Francisco harbors. Only graduate James E. Hair served aboard a combat ship, and he joined the USS *Mason* as its single black officer in July 1945 only after the conclusion of its Atlantic convoy-escort duties. Because their assignments involved relatively unimportant duties, their actual accomplishments were small. However, the thirteen were indeed "Golden," for they opened the door for black naval officers. Even though the navy allowed none of the twelve commissioned officers or the warrant officer to use the officers' clubs at their assignments, never again would the U.S. Navy's commissioned ranks be all white.

The V-12 did not commission its first black officers until after the graduation of the Golden Thirteen and never accepted more than a token representation of black students. Only fifty-two African Americans received commissions through the V-12 program, compared to seventy thousand white officers. Although most of the black V-12 graduates did serve overseas, only one advanced to the rank of full lieutenant, the rest remaining ensigns or lieutenants junior grade.

Nevertheless the V-12 graduates made their mark. On May 2, 1945, Ens. Samuel L. Gravely, of Richmond, Virginia, reported for duty aboard the USS *PC-1264* as the first black naval officer to serve on a combat ship. The white officers and the black crewmen welcomed Gravely, and on the following day he accomplished another first: the first black officer to become a member of a ship's wardroom. For the first time in U.S. Navy history, a black man was served a meal in the officers' mess rather than being the person who served it.

Although the Golden Thirteen and the V-12 graduates appropriately receive the most attention in studies of black World War II naval officers, one other instructive instance deserves mention. Early in the war the navy and army competed to enlist a limited number of doctors and other medical personnel available to serve. To win the competition with the army, the navy began offering commissions to medical students, who would assume their duties as officers upon graduation.

On June 18, 1942, the navy awarded a reserve commission to Harvard University medical student Bernard W. Robinson. Either the recruiting officer never saw Robinson or mistook the light-skinned black student for white. An internal Bureau of Naval Personnel memo later explained Robinson's commissioning as the result of "a slip by the officer who signed up medical students." The same memo also reported that the recruiting officer "says this boy has a year to go in medical school and hopes they can get rid of him somehow by then."

The recruiting officer proved to be wrong again. The "boy" did graduate and served as a U.S. Naval Reserve officer and doctor. Since he did not actually report for active duty and don his officer's uniform until after the commissioning of the Golden Thirteen, Robinson remains only an interesting footnote in black naval history rather than being the "first" black naval officer.

Like Robinson, the Golden Thirteen and the V-12 graduates received their commissions in the U.S. Naval Reserve. The primary source of Regular Navy officers continued to be the U.S. Naval Academy, and it remained all white until the final months of the war. In June 1945, Rep. Adam Clayton Powell of New York appointed Wesley A. Brown of Washington, D.C., to the academy. Despite severe harassment from his classmates, Brown remained at Annapolis and in June 1949, nearly four years after the end of World War II, graduated as the first black and the 20,699th midshipman to earn his commission from the academy.

Black officers and black-manned ships served most of World War II in segregated conditions. It was not until the conflict's last year that African Americans were given the opportunity to prove that they could function aboard integrated ships. On April 28, 1944, Secretary of Navy Knox, who had remained reluctant to support any advancement by blacks, died, and President Roosevelt appointed James V. Forrestal as his replacement.

Forrestal, a longtime member of the Urban League, brought to the office his belief in equality for all Americans as well as the practical notion that segregation had an adverse impact on morale and proved economically impractical. In July 1944, Forrestal integrated advanced training schools and a month later announced that black sailors would be assigned as 10 percent of the general crews of twenty-five selected auxiliary vessels, including ammunition ships, cargo carriers, and oilers. In a memorandum to the president, Forrestal explained that if the experiment proved successful, he planned "to extend the use of Negroes in small numbers, to other types of ships as necessity indicates."

The secretary's decision in practice applied to an extremely small portion of the seventy-five-thousand-vessel fleet because only 2 percent were auxiliary vessels and less than 2 percent of those ships were open to black sailors. Yet potentially his decision affected the whole navy.

The experiment with limited integration of the selected auxiliary-ship crews proved extremely successful. White captains reported that black sailors performed their duties well and that white crews generally accepted them. Satisfied with the experiment and facing continual shortages of sailors for the expanding fleet, the navy, in April 1945, opened all sixteen hundred auxiliary ships to black sailors, the only restriction remaining the 10 percent limitation on their numbers aboard each vessel.

Two months later, the navy announced the end of segregated basic training. Beginning in June 1945, all enlistees reported to the nearest training center regardless of their race.

Despite the advancements in equality during the conflict's final months, few African-American sailors in World War II actually received the opportunity either to train as officers or to serve aboard the two black-manned combat ships or with the auxiliary fleet. The vast majority labored in ammunition handling or construction units. Throughout the war, more black sailors served as mess attendants than in any other specialty.

African Americans did not always willingly accept the lack of opportunity and discrimination. On the Pacific island of Guam in December 1944 black sailors rioted against racial harassment from white marines. In March 1945 members of a black construction battalion at Port Hueneme, California, protested nonviolently against their white commander's racism by refusing to eat for two days.

The most significant protest by black sailors took place after the huge explosion of two ammunition ships at the Port Chicago Ammunition Depot at Mare Island, California, on July 17, 1944, which killed more than two hundred ammunition handlers, mostly black sailors assigned to a segregated unit. When loading operations resumed aboard new ships a few days later, 258 black sailors refused to return to work, citing inadequate training and a lack of safety provisions for the hazardous duty.

After discussions with military chaplains and threats from their white officers, all but forty-four returned to duty. These forty-four, along with six others, who quit loading a few days later, faced courts-martial for mutiny; they received sentences to hard labor for eight to fifteen years and dishonorable discharges. Not until after the war did lobbying by attorney and future Supreme Court justice Thurgood Marshall and editorials in the black press influence the navy to set aside the convictions and return the men to duty in January 1946.

Black women faced even more barriers to serving in the navy than did black men. When the navy organized Women Accepted for Volunteer Emergency Service (WAVES) in 1942, it excluded black applicants. Secretary Knox responded to protests from black organizations by saying that there was no need for black women in the WAVES because of the organization's stated mission. According to Knox, the WAVES were to replace men in shore-duty assignments to free them for sea duty. Since few blacks went to sea, Knox explained, there was no need for women to take their jobs onshore.

In response to the rationale of "no need" for black women in the navy, African-American leaders turned to the other primary factor that had always provided opportunities for their race in the military. After briefings and encouragement from black leaders, Thomas E. Dewey, governor of New York and Roosevelt's 1944 presidential-election opponent, made integration of the WAVES a political issue. Threatened that Dewey might win the black vote, Roosevelt directed the navy to take another look at their policy.

On October 19, 1944, just three weeks before the election, the navy announced that the WAVES would begin accepting black enlisted women and officer candidates. Plans called for segregated training for the black WAVES, but because so few volunteered due to the navy's past discrimination and the war's imminent end, black women shared classes with white recruits. Although the training of the WAVES went well, the navy did not enthusiastically recruit black women, nor did black women readily volunteer. By the end of the war, the WAVES contained a total of seventy-eight thousand enlisted women and eight thousand officers. Of these, only two officers and seventy enlisted women were black.

The Marine Corps proved even more reluctant than the other services to accept either black men or women. As a part of the navy, the Marine Corps participated in the General Board that made recommendations for the acceptance of African Americans into both services. On January 23, 1942, Gen. Holcomb, commandant of the corps testified to the board that "there would be a definite loss of efficiency in the Marine Corps if we have to take Negroes. . . ."

Holcomb openly resisted black enlistment until Secretary Knox's announcement on April 7, 1942, that the navy and marines would accept them. On May 20 the Department of the Navy announced that the Marine Corps must begin recruiting nine hundred blacks per month. According to the corps' own history of African Americans in their service, written thirty years after the war, "there is no question that the order was unpopular at Headquarters Marine Corps."

The marines trained their carefully screened recruits at New River (later renamed Camp Lejeune), North Carolina, in segregated facilities known as Montford Point. Members of the first training cycle included transfers from the navy's mess-attendants branch and several veterans of the black army regiments. White drill instructors made the training as tough for these trainees as for the rest of the corps. After the first training class graduated, several of the prior servicemen remained at Montford Point as assistant instructors, and many advanced to full drill instructors during later cycles.

The other graduates also stayed at Montford Point, assigned to the Fifty-first Defense Battalion. In February 1944 the Fifty-first, composed of field and air defense artillery units, sailed for the Pacific, where they defended installations first in the Ellice and then in the Marshall Islands. Neither of these island groups ever came under air or sea attack from the Japanese, and the first black marine unit saw no combat.

Using the experienced cadre at Montford Point that had trained the Fifty-first Battalion, the Fifty-second Defense Battalion quickly took shape and in October 1944 also sailed for the Pacific. The Fifty-second prepared defenses in the Marshalls and then served on Guam, Eniwetok, and Kwajalein. Like the Fifty-first, however, these assignments came after the Allies had already secured the islands. Prepared to defend against counterattacks that never came, the Fifty-second also saw no combat.

Ironically, while the black marine units trained for combat never actually engaged in fighting the enemy, the black service units, organized merely to support forward units, fought the enemy, suffering the only casualties by black marines during the war. The Marine Corps had formed fifty-one depot companies consisting of 110 to 165 men and twelve ammunition companies consisting of 255 men each. The duties of these black companies, which began arriving in the Pacific in the spring of 1944, were the loading and unloading of supplies and ammunition and moving them inland during island-hopping campaigns. After delivering their loads, black marines often carried the wounded back to the beach for evacuation.

During the various island operations, black marine service units frequently came under fire; nine individuals were killed in action, and another seventy-eight were wounded. Reports of the brave performance of the service units in combat on the island of Saipan on June 15, 1944, so impressed newly appointed commandant of the corps Alexander A. Vandegrift, that he stated, "The Negro marines are no longer on trial. They are marines."

Despite the praise of the commandant and the general acceptance of blacks by their fellow white marines in the combat zones, the corps made no effort during World War II to integrate its units or to reach the stated goal of 10 percent of its personnel being African Americans. In fact, fewer then twenty thousand blacks became marines during the war, less than 5 percent of the corps' total. More than 65 percent of this number served overseas, but few actually saw combat. Although a few black marines entered officers training in the war's final months, none received commissions before its conclusion. The first black marine officer, Frederick C. Branch of Hamlet, North Carolina, did not pin on his second-lieutenant bars until November 10, 1945.

Overall, the black experience in the Marine Corps during World War II was limited. Nevertheless, black marines, like those African Americans in the other U.S. armed forces, had struck down barriers and opened doors for future service. Blacks soldiers, airmen, sailors, and marines in World War II fought for the right to serve and fight. Their struggle for integration and equality would not end with the peace. In the post–World War II era, blacks would demand to maintain and increase their role in the military, which they had earned during the long conflict.

13

EXECUTIVE ORDER 9981: BEFORE AND AFTER

The more than 1 million blacks who served in uniform during World War II returned home to communities in which racism, prejudice, and discrimination had changed little. The American Legion, the Veterans of Foreign Wars, and the Disabled American Veterans all restricted black membership to segregated chapters. Due to housing segregation, few black veterans were able to take advantage of GI mortgages, and those who wanted to use the GI Bill to further their education found most colleges still closed to them. Racism, particularly in the South, also excluded a large percentage of black veterans from the many postwar job-training programs.

In 1946 mobs in the South lynched six blacks and beat numerous veterans, some still in uniform, for violating Jim Crow laws that restricted their access to public facilities. During the same year, Sgt. Isaac Woodard, recently discharged from Fort Gordon, Georgia, and still in his uniform, boarded a bus to take him to his North Carolina home. At a rest stop in South Carolina the bus driver became angry that Woodard took so long to use the "colored only" toilet facilities and called the local sheriff to arrest the black veteran. During the arrest the sheriff beat the nonresisting Woodard with a blackjack and struck him in the eyes with a nightstick, blinding him.

The military services also began reverting to their prewar positions restricting black enlistments and maintaining segregation within the ranks. Advances in equality, achieved by the lobbying of black leaders and by the contributions of black servicemen and women to the war effort began to erode with the onslaught of peace and the rapid demobilization of the armed forces.

Despite the adverse reception of returning black veterans and the tenacious discrimination throughout the United States, African Americans emerged from World War II more united than ever before. Blacks were prepared to pursue activism in peace to achieve the equality denied them during war. This unity, and support from segments of the white population, enabled African Americans to politically force the military to establish regulations designed to advance blacks in uniform. Legal decisions also began to underscore the rights of blacks to true equality. These political and legal advances made during the first five years after the end of World War II, however, were not rapid or easy in practice. Blacks still had many obstacles to overcome.

Even while racism and discrimination prevailed, more and more white Americans realized that as one of the world's most powerful countries and the symbol of freedom and opportunity, the United States could not perpetuate its discriminatory practices against more than 10 percent of its own population strictly because of their race. Members of the U.S. Supreme Court were among those forced to review the dilemma of segregation and limited opportunities. They began making decisions accordingly during the first two years of peace by ruling in favor of blacks on issues such as fair housing practices and the integration of public schools.

Military officials were also aware that African Americans remaining in uniform would not willingly return to their limited enlistment quotas and restricted jobs. Senior officers knew, too, that for the first time ever during peace the military needed black enlistees to help deter the threat of the cold war between the United States and the Soviet Union. The United States had to maintain a powerful, well-trained force.

The services had begun studying postwar employment of blacks even before World War II concluded. On October 1, 1945, a group of senior army officers assembled in Washington on the verbal orders of newly appointed Secretary of War Robert P. Patterson to study black participation in the war and to prepare a comprehensive policy on the use of black manpower in the peacetime army.

Unofficially adopting the name of its senior member, Lt. Gen. Alvin C. Gillem Jr., the Gillem Board met for six weeks, reviewing a multitude of studies and interviewing sixty military and civilian leaders. The complete report, published as War Department Circular No. 124, "Utilization of Negro Manpower in the Postwar Army Policy," and dated April 27, 1946, took steps in the right direction, but its recommendations were far from what the black community wanted and deserved. An introductory paragraph announced: "Negro manpower in the postwar army will be utilized on a broader professional scale than has obtained heretofore. The development of leaders and specialists based on individual merit and ability, to meet effectively the requirements of an expanded war army will be accomplished through the medium of installations and organizations. Groupings of Negro units with white units in composite organizations will be accepted policy."

In the implementation paragraphs that followed, the report recommended limiting black enlistments in the army to the "1 to 10 ratio of the Negro civilian population to the total population of the nation." The report, acknowledging the success of assigning black combat platoons to white units after the Battle of the Bulge in 1945, recommended that the procedure continue. No longer would divisions be all black; rather, the army would assign black platoons to white companies, black companies to white battalions, and black battalions to white regiments. The policy also clearly stated that this limited integration of African-American units into white ones did not extend into living or eating facilities, which would remain segregated.

Reactions to the policy were mixed. White civilians, enjoying the euphoria of peace, paid little attention to military affairs, and white army leaders remained so engrossed in postwar reorganization that they saw implementation of the policy as just another task in a long list of many to accomplish.

Black opinion ranged from finding the policy unacceptable to recognizing it as progress. Roy Wilkins wrote in the April 26, 1946, edition of the *Pittsburgh Courier*, "The basic policy is still Jim Crow units. Instead of having big Jim Crow units . . . we are to have nothing larger than Jim Crow regiments." A March 9 editorial in the *New York Age* noted, "The policy is still a little foggy and falls far short of its advanced advertising that it would abolish segregation in the army."

Other members of the black press supported the policy. The *New York Amsterdam News* declared on March 9 that it found the report "still some distance from the elimination of a Jim Crow army; but it represents advance and progress. As such, we say hurray and good deal."

That same day, the *Norfolk Journal and Guide* wrote, "It appears that the War Department has definitely turned the corner in policy."

Still another March 9 editorial in the *Baltimore Afro-American* declared, "We believe that the army is headed in the right direction and recommend that its sister services also get in step."

The navy was indeed already "in step" as far as studying the use of blacks in the postwar fleet, but their conservative recommendations did not go even as far as the limited advancements suggested by the army. During the last few months of World War II and on into the postwar period, the navy studied both current and future possibilities for black sailors.

Secretary Forrestal appointed Lester B. Granger, an Urban League official, as his civilian adviser on black policies in March 1945. Granger immediately set out on a fifty-thousand-mile tour of ships and facilities to survey current attitudes and practices among white and black naval personnel. He found a great range in the degree of integration, with the amount of segregation depending on the location and commander. The most consistent finding by Granger at all locations was that the greater the degree of segregation, the lower the morale and proficiency of all on base.

Forrestal, correlating Granger's recommendations with his own experience, had initiated measures to integrate the navy even before the end of World War II. In June 1945 the navy had abolished separate training facilities for black recruits and integrated all navy schools. The following February 27, while the navy was in the midst of downsizing its personnel and fleet, Forrestal had directed that "all restrictions governing types of assignments for which Negro naval personnel are eligible are hereby lifted."

Meanwhile, the Marine Corps, which had only begrudgingly accepted their first black recruits during the war, quickly initiated measures to reduce those numbers with the arrival of peace. Immediately after the surrender of Japan, the corps announced plans to retain only

2,880 black marines in a postwar force of 100,000—less than three percent of the corps' total—and they would remain segregated and restricted to antiaircraft units and support forces. Then, in early 1947, the Marine Corps amended its policy, allowing only black stewards, a specialty equivalent to the navy messman, to remain on active duty.

The Marine Corps' position came from the long-standing attitude that the military should be the country's defense structure, not a social experiment. In a May 28, 1946, memo, headquarters of the Marine Corps explained their position:

> It appears that the Negro question is a national issue which grows more controversial yet is more evaded as time goes by. During the past war the services were forced to bear the responsibilities of the problem, the solutions of which were often intended more to appease the Negro press and other "interested" agencies than to satisfy their own needs. It is true that a solution to the issue was, and is, to entirely eliminate any racial discrimination within the services, and to remove such practices as separate Negro units, ceilings on the number of Negroes in the respective services, etc., but it certainly appears that until the matter is settled on a higher level, the services are not required to go further than that which is already custom.

The Marine Corps made it obvious that they were not going to take the lead in integration and the pursuit of equality. As harsh as their stance appears, it was perhaps more honest than those of the army and the navy. Each of these services professed to be moving toward integration and equal opportunity for all; in fact, however, the rights of black soldiers and sailors evaporated in the rhetoric white military leaders used instead of action to end segregation.

The army and the navy strictly enforced the limitation of enlisting only enough blacks to maintain a level of 10 percent of the force. Acceptance standards remained so strenuous, however, that the numbers of blacks in uniform actually began to fall rather than climb toward the 10 percent quota.

Major organizational changes occurred in the military during the late 1940s, but they proved to have little effect in increasing opportunities for blacks. In September 1947, Forrestal assumed leadership of the newly formed Department of Defense, which supervised the redesignated Department of the Army, the Department of the Navy, and the former air arm of the army that had become a separate and equal Department of the Air Force.

The navy still controlled the Marine Corps, which openly resisted retaining African Americans. The now-independent air force concentrated its black enlisted men at a few southern bases, where they performed mostly laborer and housekeeping chores. With closure of the Tuskegee Air Field, all black pilots were transferred to the 332nd Fighter Group, commanded by Col. Benjamin O. Davis Jr., at Lockborne Field, Ohio. By 1948 blacks represented less than 3 percent of the marines and 6 percent of the airmen on active duty.

At the same time, the numbers of blacks in the navy had fallen to less than 5 percent, and nearly two-thirds of them remained messmen. A black man in the postwar navy, despite integrated training and promises of equality, was still more likely to be a servant than a regular sailor.

The army assigned black soldiers to segregated organizations and slowly infused those units into white commands. On many army posts, black soldiers picked up trash, cut the lawns, and performed other manual-labor duties rather than train for combat. The rare opportunities for blacks in the army led to few volunteers, and by the summer of 1947 less than 9 percent of soldiers were African Americans.

Black leaders and newspapers, recognizing that the military was more talk than action, persevered with their demands for equal rights in the ranks. The military paid no attention to black protests and even less to their own equality policies. In the armed forces, it was business as usual.

With pressing need no longer the impetus for change in a reluctant military, politics took the forefront. President Harry Truman, having been reared in Missouri and retaining many of the racial prejudices of his generation—including referring to "niggers" in private conversation with fellow whites—nevertheless possessed a strong sense of justice. Despite his upbringing, Truman became one of the first politicians to declare his support of equality for all races. In a June 15, 1940, speech at Sedalia, Missouri, while he campaigned for the Senate, Truman declared, "In giving Negroes the rights that are theirs, we are only acting in accord with our ideals of a true democracy."

Truman acknowledged that blacks had willingly served in time of war and now deserved to enjoy the benefits of the peace. However, the president was first and foremost a politician and, like most elected leaders, tended to take only stances that ensured winning elections. Although concerned about discrimination against blacks both in and out of the military, Truman did little as president to right past wrongs until it became politically expedient.

When the Republicans gained control of Congress in the 1946 election, the Democratic Party, no longer confident that Truman could be elected in 1948, splintered. Henry A. Wallace, once Roosevelt's vice president, formed the Progressive Party from liberal Democrats who opposed the president's hard line against the Soviet Union. Southern Democrats, angry at Truman's pro-equality leanings, bolted and formed the States' Rights Democratic Party. As "Dixiecrats," they nominated Gov. Strom Thurmond of South Carolina as their presidential candidate and adopted a platform based on continued segregation.

Adding further pressure on Truman were black leaders, who wanted him to take a firmer stand on integration of the armed forces. The president appointed committees to study general civil rights as well as the future of blacks in the military, but African-American leaders were tired of discussions that preempted action. On November 23, 1947, A. Philip Randolph and other prominent black leaders established the Committee Against Jim Crow in Military Service and Training and encouraged civil disobedience if discrimination and segregation in the draft and in the armed services did not cease. Over the next few months, the NAACP, as well as other black organizations, endorsed the committee's demands.

With his own party split three ways, Truman and his advisers were aware that the key to possible victory was the influential black vote. Yet opposition from black leaders and their white supporters to segregation still did not move Truman to take action.

The final blow to any apparent possibility of Truman's winning the 1948 election came when the Republicans nominated Thomas E. Dewey and resolved in their campaign platform: "We are opposed to the idea of racial segregation in the armed forces of the United States."

The Republican announcement forced Truman to take strong and immediate action to win the black vote and the election. He decided that even if a bold stroke did not get him elected, he at least could do the "right thing" and conclude his presidency in an honorable way. On July 26, 1948, President Truman issued Executive Order 9981, which declared:

> It is essential that there be maintained in the armed service of the United States the highest standards of democracy, with equality of treatment and opportunity for all those who serve in our country's defense. It is hereby declared to be the policy of the President that there shall be equality of treatment and opportunity for all persons in the armed forces without regard to race, color, religion, or national origin. This policy shall be put into effect as rapidly as possible, having due regard to the time required to effectuate any necessary changes without impairing efficiency or morale.

Immediately controversy erupted over what Executive Order 9981 really meant. Some thought the order officially ended discrimination but not segregation, while others read the words as the end of segregation.

Three days after announcing the order, Truman ended the confusion at a press conference. A reporter asked, "Does your advocacy of equality of treatment and opportunity in the armed services envision eventually the end of segregation?" Truman provided a one-word, easy-to-understand answer: "Yes."

Over the next few weeks A. Philip Randolph and other black leaders announced their support of Truman, but the president still entered the November election as the underdog. However, the black vote, particularly Ohio, Illinois, and California, provided an upset victory for the president.

In the months between Truman's signing Executive Order 9981 and his election, neither political nor military leaders had taken any significant action, believing that Truman would be defeated and the policy never implemented. Truman's victory produced action. The president's seven-man Committee on Equality of Treatment and Opportunity in the Armed Forces held its first meeting on January 12, 1949, as authorized by his executive order, "to determine in what respect such rules, procedures, and practices may be altered or improved with a view to carrying out the policy."

With a flourish of his pen, Truman had ended segregation in the armed forces, but translating words on paper into actions could not, and would not, be a miraculous process. Many military leaders, out of concern for their service's efficiency, or racism, or a mixture of both, still opposed integration. They would follow orders, but they would do so at a deliberate pace.

Truman appointed Charles Fahy, the Georgia-born former U.S. solicitor general and diplomat, to head the committee that for all but official purposes took his name. After their first meeting, the Fahy Committee reported to Truman that they preferred to negotiate with the individual services to reach amiable agreements rather than issuing a direct order implementing the executive order. Truman agreed.

Officially, the Fahy Committee acted as an advisory board to the secretary of defense, but the service chiefs recognized that it represented the president's objective of integrating the ranks. During the next two years, the committee proceeded in a spirit of understanding and cooperation while at the same time prodding and prompting the services to provide opportunities for blacks and to end segregation.

The committee's successes were neither instant nor all-consuming, but the group did push the services to take steps in the direction of overcoming more than 175 years of discrimination and racism. From its opening session, the Fahy Committee found the navy and air force fairly willing to accept change, while the army and marines resisted integration. While acknowledging the navy's progress in opening the general service to black recruits, the committee members expressed concern about the overall number of African-American seamen, especially the few officers, on active duty and encouraged the increase of recruitment efforts. It also recommended that the navy make additional efforts to move more black stewards out of the dining rooms and galleys into the other specialties and to allow those remaining as messmen to be promoted to the rank of petty officer. On June 23, 1949, the navy adopted all of the committee's recommendations.

The air force received permission from the committee to develop their own plan to institute the executive order. On May 11, 1949, the Department of the Air Force announced "that there shall be equality of treatment and opportunity of all persons in the air force without regard to race, color, religion, or national origin." The air force also ended quotas on black enlistment, assignment, and promotion by declaring that it would base all future actions on merit and ability, not on race.

As its first step toward integrating all units, the air force disbanded the all-black 332nd Fighter Wing and reassigned its pilots to previously all-white squadrons or to schools for further training. Black enlistments began to increase by more than five hundred men per month, and over the next two years the air force reduced its number of segregated units by half. By 1950 all air-force jobs and schools were without racial quotas or restrictions, and all black airmen in units and schools lived in integrated conditions.

The marines were another story. Gen. Clifton D. Cates, the commandant of the Marine Corps, responded on March 19, 1949, that the military could not "be an agency for experimentation in civil liberty without detriment to its ability to maintain the efficiency and high state of readiness so essential to national defense."

Cates expressed sympathies with the ideas of equality and integration but concluded with an opinion shared by many senior military officers:

> The problem of segregation is not the responsibility of the armed forces but is a problem of the nation. Changing national policy in this respect through the armed forces is a dangerous path to pursue in as much as it affects the ability of the National Military Establishment to fulfill its mission. . . . Should the time arise that non-segregation, and this term applies to white as well as Negro, is accepted as a custom of the nation, this policy can be adopted without detriment by the National Military Establishment.

Regardless of his strong feelings, Cates had no recourse but to honor the order of President Truman, his commander in chief, and the navy's policy, issued on June 23, 1949, supporting racial equality, which also extended to the Marine Corps. On November 18, 1949, the marines issued a memorandum of guidance to subordinate commanders revoking all previous policies prohibiting integration. The memorandum did not disband current all-black units but did direct that any marine, regardless of race, could be assigned to any unit in any specialty. Effectively, the order ended segregated training at Montford Point and paved the way for integrating the corps.

Of all the services, the army provided the strongest and most tenacious opposition to Executive Order 9981 and the Fahy Committee. Secretary of the Army Kenneth C. Royall told the committee in March

1949 that the army was "not an instrument for social evolution" and that integration would undermine the army's efficiency. He reported that during the past two world wars blacks had failed to prove themselves in combat and that he believed them qualified only for laborer duties. In pointing out to the committee that the army had more black officers than the other services, the secretary said he would work to increase black recruitment, but he remained steadfast that there was no place for general integration in the army.

A few weeks after his report to the committee, Royall resigned, and Gordan Gray, a North Carolina lawyer, became the new secretary. In his attempts to institute the president's directive, Gray found the army's senior officers resistant to changing current segregation policies and enlistment quotas. Fahy finally had to secure direct support from Truman to force the army to comply. On September 30, 1949, the army opened all schools and specialties to blacks. The following January 16, the army approved integration for all units and on March 27, 1950, ended quotas for black enlistments.

In June 1950 the Fahy Committee made its progress report to the president and to the American people. *Time*, in its June 5 edition, provided a simple explanation for the work of the committee, stating: "The committee did not argue the moral or sociological aspects of the case. It based its arguments on efficiency. There were bright Negroes and there were dumb ones, just like white men. To refuse a job to an intelligent or skilled Negro was simply a waste of manpower. Concentration of unskilled Negroes in segregated units just multiplied their inefficiency."

The *Time* article was not a resounding endorsement of integration or of equal rights, but it did note the formal end of segregation in the services. However, no presidential order or committee finding could end racism and discrimination. Although the services now had official policies ending segregation and racial restrictions, the implementation by whites was slow and mostly unenthusiastic. Politics had put the policies in the books, but only the needs of war could convert regulations into action.

14

THE KOREAN WAR: INTEGRATION UNDER FIRE

On June 25, 1950, more than 100,000 North Korean soldiers, supported by fourteen hundred artillery pieces and 125 Soviet T-34, tanks crossed the 38th Parallel and moved into South Korea. The North Koreans routed the South Koreans and the few U.S. military advisers who had remained in the country after the bulk of the World War II occupation forces had withdrawn to Japan or to the United States.

President Truman committed American ground and air troops to stem the invasion supported by a U.S. anti-Communist policy and backed by a UN Security Council resolution. This act swiftly transformed integration of the armed forces from a political issue to a non-issue as requirements for personnel escalated.

At the time of the North Korean attack, implementation of integration policies was plodding slowly forward and encountering delays and resistance, especially from the army and Marine Corps. These problems paled, however, compared to the overall lack of U.S. military preparedness for the battle on the Korean Peninsula only five years after the conclusion of World War II. Most of the military's focus had been on halting the spread of communism in Europe and preparing for a possible conflict with the Soviet Union. American units assigned to Japan and the other Pacific installations were poorly trained and equipped, prepared neither mentally nor physically to face the aggressive North Koreans.

In Japan, white and black servicemen enjoyed the comforts of an occupation army in which low-paid Japanese servants cleaned their barracks and even shined their shoes. The American soldiers were far from combat ready. Yet less than three weeks after the North Korean invasion, these troops joined units from other Pacific islands to form four army infantry divisions and a marine brigade. Their mission was to turn back the Communist offensive, but unable to do so, they withdrew into a tight perimeter, seventy miles wide by sixty miles long, around the coastal city of Pusan.

Despite repeated acts of individual bravery during these early battles, the overall inefficiency of the units overshadowed any heroism. Companies and battalions broke and ran to the rear in the face of enemy assaults. "Bugging out" by unprepared, poorly trained units was common across the Korean front, but no unit faced more negative publicity and condemnation for their poor performance under fire than did the Twenty-fourth Infantry Regiment.

Integration had not reached the Twenty-fourth Infantry in Japan prior to its deployment to Korea, and all the cumulative effects of segregation were prominent in the unit. African-American enlistees assigned to the Twenty-fourth, like other black soldiers, had endured an inferior segregated education system in the United States which resulted in their consistently scoring lower than whites on military entrance tests. Because those who scored in the lowest quadrant on the aptitude test received infantry assignments to segregated units, the Twenty-fourth had more than its share of undereducated troops. Another holdover effect of segregation for the unit was that white officers still viewed assignment to black regiments as detrimental to their careers. Thus, white officers in black units tended to be the young, the inexperienced, and the incompetent. And finally, as a segregated unit, the Twenty-fourth suffered from a low priority in terms of opportunity for combat training.

Manned by soldiers with low marks on the army's aptitude tests, led by officers who were marginally qualified, and with little or no combat training over the past year, the Twenty-fourth Infantry entered the Korean War as ill prepared for warfare as any U.S. unit in history—white or black. Joining two white regiments as a part of the Twenty-fifth Infantry Division, the Twenty-fourth Regiment engaged the North Koreans almost as soon as it landed on the penin-

Gun jeeps of the Twenty-fourth Infantry Regiment in Korea, 1950. (U.S. ARMY MILITARY HISTORY INSTITUTE)

sula. The result was disastrous. In an early fight most of the enlisted men in one rifle company melted away and left the unit's officers and black sergeants to man their sector. A few days later, most of a battalion withdrew without orders, exposing several artillery batteries to direct enemy attack.

White units experienced similar difficulties, but these problems diminished as the senior headquarters assigned new commanders, replaced the unfit troops, and reorganized the forces to add to their confidence and fighting ability. Similar changes were impossible in the segregated Twenty-fourth because no black replacements or units were available to reinforce or to replace ineffective soldiers.

As the U.S. and remaining South Korean forces withdrew into the Pusan Perimeter, more and more stories circulated about the "bug-out Twenty-fourth." White soldiers and commanders in units who had performed just as poorly welcomed the opportunity to heap abuse on the black unit to distract attention from their own failures. Racists, both within and outside the military, cited the problems in the Twenty-fourth Regiment to justify their stance on the inferiority of African Americans as they repeated the jingle "When them mortars begins to thud, the old deuce-four begins to bug."

While the Twenty-fourth had not performed well as a unit, many of its soldiers had fought magnificently. In July 1950, Texas-born second lieutenant William D. Ware earned the Distinguished Service Cross for gallantry near Sangju. According to the award citation, "the position was attacked from three sides by numerically superior enemy forces armed with automatic weapons and supported by mortar fire. . . . Lieutenant Ware, arming himself with a rifle, ordered his men to withdraw. He was last seen firing . . . on the advancing enemy until his position was overrun."

A month later, near Haman, Korea, on August 2, 1950, PFC William Thompson, of Company M, Twenty-fourth Infantry Regiment, displayed individual bravery not uncommon to black soldiers in Korea. Thompson, an orphan raised in the New York Home for Homeless Boys, had been in Korea for only five weeks when his platoon came under a surprise night attack by a far larger enemy force. The twenty-two-year old African-American soldier moved his machine gun to the center of the enemy advance and directed withering fire upon the advancing enemy, sufficiently slowing their attack so that his platoon could withdraw to more defensible positions.

Although wounded by small-arms fire and grenade fragments, Thompson stayed behind his machine gun. As the remaining few members of his platoon withdrew, a sergeant ordered Thompson to join them. The young machine gunner refused and continued to provide covering fire. "Maybe I won't get out, but I'm gonna take a lot of 'em with me" were his last words before an exploding enemy hand grenade killed him.

Another Twenty-fourth Infantry soldier, Cornelius H. Charlton of Company C, advanced to the rank of sergeant after eight months of combat in Korea and earned his regiment's second Medal of Honor of the war near Chipo-ri on June 2, 1951. Charlton, from an East Gulf, West Virginia, family of seventeen children, assumed command of his platoon when the unit's white officer was evacuated with wounds received in an attack on Hill 542. Advancing at the front of his platoon, the twenty-one-year old African-American sergeant killed six enemy soldiers and destroyed two of their positions with his rifle and hand grenades.

Charlton continued up the hill, only to be halted by a shower of grenades and small-arms fire. Despite severe shrapnel wounds to the chest, Charlton refused medical treatment, regrouped his men, evacuated the wounded, and made still another assault against the entrenched enemy. When the platoon reached the top of the hill, they came under fire from still another enemy strong point on the reverse slope. Charlton, in a one-man assault, charged the position and killed the remaining enemy soldiers. His Medal of Honor citation concluded: "The wounds received during his daring exploits resulted in his death, but his indomitable courage, superb leadership, and gallant self-sacrifice reflect the highest credit upon himself, the infantry, and the military service."

Individual bravery was not, however, sufficient to change the reputation of the Twenty-fourth Infantry Regiment as a unit that consistently broke under fire. The commander of the Twenty-fifth Infantry Division, Maj. Gen. William B. Kean, claimed that the Twenty-fourth Regiment was his weakest subordinate element. Kean added that the black regiment not only endangered his division but also weakened the entire UN effort against the North Koreans. As an immediate solution, Kean recommended that the Twenty-fourth Regiment be disbanded and its soldiers integrated into the division's white units using a one-to-ten ratio.

In the midst of the initial victories by the North Koreans that threatened to push the Americans out of Pusan in a Dunkirk-like evacuation, all the U.S. command could do was to continue to commit replacements and begin a protracted air war against the enemy's rear areas and resupply lines. Finally, the Americans and UN forces slowed and then stopped the North Korean advance as the Pusan perimeter held firm. In September, less than three months after the invasion, Gen. Douglas MacArthur, in command of all UN troops, conducted an amphibious assault on the enemy's rear at Inchon. A few days later, the forces in the Pusan perimeter broke out and began pushing the North Koreans back across the 38th parallel.

The Americans tended to forget their initial defeats as the North Korean army began to disintegrate during its retreat northward. The lasting memory of many, however, was the bug-out reputation of the Twenty-fourth Regiment.

Meanwhile, back in the United States, the prewar integration policies and the nearly overwhelming numbers of recruits reporting for duty were ending segregation in the army's training centers. At Fort Jackson, South Carolina, training center commander Brig. Gen. Frank McConnell attempted in the spring of 1950 to establish separate units for black and white recruits, but he soon found it impractical to separate by race the one thousand daily arrivals.

Citing Secretary of the Army Gray's integration order of January 16, 1950, McConnell informed his senior commanders of his intentions and then with a verbal order ended segregated training at Fort Jackson. Other than a few editorials in southern newspapers opposing the change, integrated training proceeded with no problems. McConnell explained the program's success in a later interview, declaring, "I would see recruits, Negro and white, walking down the street, all buddying together; the attitude of the southern soldiers was that this was the army way; they accepted it the same way they accepted getting booted out of bed at five-thirty in the morning."

Over the next few months news of the success of integrated training at Fort Jackson spread to the senior command in Washington and to the other training centers across the United States. By the end of the year all ten army basic-training centers had successfully integrated.

Simultaneously, in Korea some commanders were, like McConnell, making decisions on integration without direction from their senior headquarters. When the Ninth Infantry Regiment of the Second Infantry Division arrived in Korea in July 1950, it contained two understrength white battalions and a black battalion that was 10 percent over strength. Casualties and no replacements reduced the number of whites even more in the first few weeks of fighting. To balance his battalions, the regiment's commander, Col. John G. Hill, began transferring black soldiers to white units. Hill later noted that the mixed units worked well and that he never had any doubt that they would do so "because at a time like that, misery loves company."

While the Ninth Infantry Regiment and other early-arriving units of the Eighth Army in Korea integrated their units when circumstances required, the overall commander of U.S. and UN troops did not support the practice. General MacArthur had underutilized black units in his Pacific command during World War II and showed no interest in integrating his command in Korea. Several observers noted that not a single black enlisted man or officer served in MacArthur's headquarters.

MacArthur's chief of staff, Lt. Gen. Edward M. Almond, shared his superior's attitudes about keeping the races separate. Almond, apparently still bitter from his lack of success while commanding the black Ninety-second Infantry Division during World War II, held firm to his prejudices when given command of X Corps, which conducted the invasion at Inchon.

When Almond arrived in Korea and found that several units in his command had already integrated, he immediately began to reverse the practice. Almond ordered that no additional blacks be assigned to white combat units and that those blacks already present not be replaced when rotated out of the units. Under Almond's directions, blacks once again dominated the rear-area service and support units.

MacArthur, by simply ignoring the issue, and Almond, by taking direct action, halted and then turned back the progress being made in integrating frontline units. Even when casualties outnumbered replacements and units were fighting far below their authorized strength, MacArthur and Almond still refused to integrate. In fact, the two senior commanders allowed untrained South Koreans to fill the gaps in the frontline commands when the personnel situation reached a critical point rather than allow integration of African Americans into the white units.

The two, however, could not hold back the tide. MacArthur complained publicly about the constraints placed on him by Washington and proposed, in clear opposition to the president's objectives, taking the war into China, if necessary, to achieve total victory. As a consequence of his insubordination, President Truman relieved the general of his command on April 11, 1951.

Gen. Matthew B. Ridgway replaced MacArthur as commander of U.S. and UN forces and immediately began plans to integrate all American units. The Virginia-born Ridgway, who opposed segregation on the moral ground that it was "both un-American and un-Christian," paused only long enough to request that the Department of the Army formally approve his desegregation efforts. He did this because of potential political difficulties with the states that provided two federalized National Guard divisions, the Fortieth Infantry from California and the Forty-fifth Infantry from Oklahoma.

Gaining quick approval from Washington, Ridgway proceeded during the summer of 1951 to integrate blacks already in Korea into white units and to do the same with black replacements from Japan and U.S. training centers. In addition to integrating units by transferring blacks and whites, Ridgway disbanded black units and transferred the troops to already integrated assignments.

Because integration of the army in Korea took place in the midst of the fighting, the change created few problems: Frontline soldiers had more than enough enemy troops to combat without turning on each other. The army further defused controversial reaction to the policy by delaying an official announcement about the action until July 21, 1951, when the implementation was already in progress. Generally, the news was well received. Blacks welcomed the overdue policy of having equal opportunity to serve in any unit in any capacity. Whites also approved. They shared the feeling that integration was long overdue or possessed the historical sense that in time of need blacks had just as much right to fight and die as did anyone else.

While Ridgway integrated his forces in Korea, the Department of the Army commissioned a study to determine the results of mixing the races in combat units and stateside training centers. Project Clear, manned by civilians from the Operations Research Office of Johns Hopkins University, interviewed enlisted men and officers of both races in Korea and the United States. The study and report of Project Clear were classified secret to limit any political use of the findings and also to prevent the disclosure of any great problems uncovered.

Ultimately the secrecy proved to be unwarranted. More than 85 percent of the officers queried said that blacks in integrated units performed as well as white soldiers. Eighty-eight percent of those interviewed stated that black officers performed as well as white officers. Evidence revealed that blacks showed "substantially the same frequencies of desirable and undesirable combat behavior" as white soldiers and concluded that "the similarities far outweigh the differences."

On November 21, 1951, the study group delivered its formal report to the Department of the Army. The findings stated conclusively that integration had increased the effectiveness of the army in Korea and in the U.S. training centers. They recommended that integration be extended to the rest of the army as soon as possible and that no quotas be enforced on the numbers of black enlistees. By the end of the war, nearly 220,000 black enlistees had entered the army and made up 12.8 percent of the its authorized strength around the world.

Backed by the results of Project Clear and the support of commanders in Korea and at the training centers, the army proceeded to integrate all of its units. Some commands were more reluctant than others, but by the spring of 1952 the entire U.S. Army was integrated. The last opposition came from Virginian Thomas C. Handy, the commander of army troops in Europe. Handy made no plans to integrate his command until directed to do so by Army Chief of Staff Gen. J. Lawton Collins. In a last measure of protest, Handy made sure that everyone knew that integration was not his idea. In an order to his command on April 1, 1952, Handy declared: "The Department of the Army has directed this command to initiate a . . . program of racial integration."

While the actual integration of the army progressed well, white commanders, many of whom were not familiar with black soldiers and some of whom still harbored racist sentiments, did not necessarily treat all their men the same. The cause of what would become a common complaint of black soldiers in the integrated army over the next decades quickly became apparent: Integration did not necessarily mean equality in terms of types and amount of punishment administered under the guise of military justice.

In late 1950 more than thirty soldiers assigned to the Twenty-fourth Infantry contacted the NAACP requesting an investigation into why black soldiers were more likely to be court-martialed and, when convicted, receive more severe sentences than white soldiers. Statistics supported the claim: Twice as many blacks were court-martialed in Korea than whites, even though they made up only about 15 percent of the total force. Black soldiers also noted that few black officers sat on court-martial boards and that trials of blacks were so expedited that they were often concluded with a guilty verdict in less than an hour.

With permission of the army, Thurgood Marshall of the NAACP spent five weeks in Korea and Japan investigating the complaints of unequal justice. Marshall found many of the soldiers' complaints valid and wrote an article entitled "Summary Justice: The Negro GI in Korea" for the May 1951 edition of the *Crisis*. In his article Marshall noted that prior to his visit to Korea, the Inspector General's Office and the Office of the Judge Advocate General in Korea were all white and that each staff unit had added one black only days before his arrival.

Marshall then focused his article on the unequal justice he found in Korea and its effect on morale. He pointed out that black soldiers convicted of "misbehavior" before the enemy received court-martial sentences of death or life in prison, while white soldiers were sentenced to prison terms of only three to five years for the same offense. Trials of black offenders seemed to focus on a rush to justice, with defense councils rarely provided time to prepare or present their cases. Most disturbing to Marshall were four court cases that lasted less than one hour each in which black soldiers received life sentences. Marshall concluded, "I have seen many miscarriages of justice in my capacity as head of the NAACP legal department. But even in Mississippi a Negro will get a trial longer than forty-two minutes if he is fortunate enough to be brought to trial."

Over the next months, the acceleration of integration and the end of all-black units took away attention from unequal military justice. Marshall's visit did little to reduce the number of blacks court-martialed or to produce a parity in sentences with white offenders. Although the events of the Korean War itself pushed the issue into the background for the time being, the questions would rise again.

The needs of war had finally made the integration policy of politics a reality in the army. By 1953 more than 90 percent of black soldiers served in integrated units, and by the end of the next year, the remaining all-black units were integrated or disbanded. The only change that brought any sadness to the black community and their supporters was the demise of the four black regiments which had proudly represented African Americans since the end of the Civil War. The Ninth Cavalry Regiment became the 509th Tank Battalion on October 20, 1950, and was integrated on March 7, 1953. Also on October 20, 1950, the Tenth Cavalry was transferred to the 510th Tank Battalion and completed integration on December 31, 1952.

Two of the battalions of the Twenty-fifth Infantry Regiment became the Ninety-fourth and Ninety-fifth Infantry Battalions before being deactivated on December 20 and 22, 1952. The regiment's Third Battalion became the integrated Twenty-fifth Armored Infantry Battalion, assigned to the First Armored Division on November 20, 1952.

The Twenty-fourth Infantry Regiment never survived the poor reputation it gained in the Korean War's first months. It was officially dis-

Soldiers of the Twenty-fourth Infantry Regiment near the Han River in Korea, 1951. (U.S. ARMY MILITARY HISTORY INSTITUTE)

banded on October 1, 1951, and replaced with the Thirty-seventh Infantry Regiment. Soldiers of the Twenty-fourth integrated into the Thirty-seventh or into other white units of the Twenty-fifth Infantry Division, and while individual soldiers continued to serve bravely in their integrated units for the remainder of the war, the Twenty-fourth's reputation remained tarnished.

Most histories of the Korean War make mention of the poor performance of the Twenty-fourth Infantry Regiment without noting any reasons or the fact that white units made the same mistakes in the confusion of the war's early months. On June 24, 1995, more than thirty years after the Korean War ended, the Department of the Army issued a press release announcing the completion of a study by the U.S. Army Center of Military History on the performance of the last all-black regiment.

In attempting to represent an accurate portrayal of the Twenty-fourth Infantry Regiment, analysts interviewed four hundred individ-

Soldiers of the Twenty-fourth Infantry Regiment on patrol near the 38th parallel in 1951. (U.S. ARMY MILITARY HISTORY INSTITUTE)

uals, two-thirds of them African Americans, and spent eight years in researching and synthesizing the information. Before its release "a number of distinguished and highly qualified individuals with particular credentials and interest in the subject" reviewed the exhaustive study and concluded:

> The story of the 24th Infantry Regiment in Korea is a difficult one, both for the veterans of the unit and for the army. The results of the study clearly suggest that what happened to the regiment was largely a product of the injustices that afflicted black soldiers prior to the formal integration of the services. In addition to the lapses of command and the deficiencies in leadership, training, and equipment that burdened this unit during the initial states of the conflict, the regiment labored under special circumstances, unique to itself.
>
> This research effort reflects the army's recognition of the many lasting effects of institutional racism which still linger from the era

of segregation despite the significant progress that has been made since then in striving for racial equality in the force.

In April 1996 the army distributed advance copies of the study, *Black Soldier, White Army*. The following month, the army announced that, still not satisfied with the study, it had delayed its publication. Whatever the report states, a white officer who led one of the regiment's platoons during the Korean War may have already recorded the last word on the subject. In the preface to his memoirs, Lyle Rishell wrote, "The black soldier can be very proud of his devotion to duty and the battles he fought in my unit in Korea."

The Korean Conflict proved to be mostly a land war, with the army and marines shouldering the bulk of the responsibility and suffering the most casualties. At the beginning of the war, the Marine Corps, like the army, remained resistant to integration and continued to restrict the number of black enlistees as well as the specialties open to them.

When North Korea crossed the 38th parallel in June 1950, the U.S. Marine Corps, 74,279 strong, numbered only 1,075 blacks, 427 of whom served as stewards. As the corps grew rapidly in response to troop requirements in Korea, the need for personnel, combined with President Truman's desegregation policy, forced the marines to accept additional black enlistees. Initially, the marines assigned the black recruits to service positions as ammunition handlers, supply clerks, and cooks. As the war progressed, however, combat losses in infantrymen and artillerymen required so many replacements that the marines began assigning blacks to frontline units.

Need and political policy were reinforced by performance; black marines proved that they could fight as well as their white comrades. Individual prejudices remained, but sharing of battlefield hardships and dangers forged white and black marines into cohesive units. In two short years the Marine Corps, under the leadership of Commandant Clifton B. Cates, advanced from the most segregated service to one that approximated total integration. Only two years into the Korean War, black marines numbered fifteen thousand and represented 6 percent of the corps. All marine facilities, specialties, and units in the United States and in the war zone were integrated.

Marines arriving in Korea found that their assignments were based on need and training rather than race. In a letter to his college newspaper that appeared in the November 4, 1952, edition of the *Daily Northwestern*, marine lieutenant Herbert M. Hart explained, "It doesn't make any difference if you are white, red, black, green, or turquoise to the men over here. No record is kept by color. When we receive a draft of men, they are assigned by name and experience only. . . . There's no way we can find out a man's color until we see him, and by that time he's already in a foxhole, an integral part of his team."

In an oral-history interview conducted by the Marine Corps Historical Division more than twenty-five years after the war, Gen. Oliver P. Smith, commander of the First Marine Division during the conflict, responded when asked if his units were integrated: "Oh, yes, I had a thousand Negroes, and we had no racial troubles. The men did whatever they were qualified to do. There were communicators, there were cooks, there were truck drivers, there were plain infantry—they did a good job because they were integrated, and they were with good people. . . . Two of these Negroes got the Navy Cross . . . and there were plenty of Silver Stars and Bronze Stars, and what have you. And I had no complaint on their performance of duty."

Perhaps most telling about the successful integration of the marines is a note in the corps' official history about African Americans joining their service. In reference to Korea, the history notes, "Records pertaining to black marines, aside from strength and deployment statistics, are virtually nonexistent. With the end of segregation, black marines merged into the mainstream of marine corps experience."

The Korean War had less effect on the integration of the navy and air force than it did on the army and Marine Corps. Unlike the latter two services, the navy and the air force entered the conflict with firmly established integration policies. The progress of integration within the navy and air force, however, differed greatly.

By late 1952 the air force had disbanded its last all-black unit and had established a policy of assigning black airmen to all stations in any capacity in which they were trained. Meanwhile, the navy continued to profess its acceptance of integration and equal opportunity for black sailors while not actually instituting all their policies and promises. Even though the navy increased in size and the number of black sailors rose from fifteen thousand to twenty-four thousand to meet requirements in Korea, their percentage of the force actually decreased to below 4 percent of the total.

At the beginning of the war, the navy still had more than 65 percent of its black sailors assigned to the messman's specialty. Despite all the policies and promises of opportunity for black sailors, fully one-half of black enlistees continued to be assigned as servants.

While the services made somewhat different progress in integrating the enlisted ranks, all were guilty of foot dragging in respect to promoting equal opportunity for blacks to become officers. Blacks who did secure commissions, however, could serve in all specialties, including as air-force pilots and naval and marine aviators. First Lt. Dayton Ragland, a black air force pilot, earned the distinction of shooting down the first North Korean jet fighter of the war. Overall, however, black officers did not even make up 1 percent of the officer corps in any of the services during the entire conflict.

These numbers were the result of the low acceptance rate of blacks into college ROTC programs and into the services' officer candidate schools. West Point and Annapolis continued to admit only a token number of black cadets and midshipmen. Between 1950 and 1955 an average of less than four black men a year received commissions from the U.S. Military Academy, and in 1956 the number dropped to a single black graduate. No black midshipman graduated from the U.S. Naval Academy in 1950 or 1951, and the classes of 1952–55 each yielded but a single black graduate.

Neither black nor white women played a significant role in the Korean War. In June 1948, Congress passed the Women's Armed Forces Integration Act, which granted women of all races a permanent place in the regular and reserve components. They remained, however, restricted to a few specialties, and their numbers were limited by law to only 2 percent of the total military strength.

Despite the expansion of the services for Korea, none of the services' women's organizations ever reached the 2 percent limit. Most served in clerical positions or as nurses. No more than a token representation of black women were inducted into any of the services during the conflict.

When the Korean War concluded with an uneasy cease-fire on July 27, 1953, the boundary between North and South Korea remained almost identical to the prewar division line. While the war had done little to change the political balance of power in the Korean Peninsula, it had expedited the desegregation of the U.S. armed forces. At the end of the war, the military was the most integrated institution in the United States. Individual prejudices and institutional racism persisted in varying degrees, but the services had indeed become sociological laboratories, proving that integration not only could work; it could work well. The great strides forward on behalf of blacks, however, would ultimately lead to more challenges in ensuring the ongoing movement toward equality in the military.

15

THE FIFTIES: AN INTEGRATED MILITARY IN A SEGREGATED AMERICA

In its February 22, 1954, edition, *Time* reported that the military was the most integrated institution in the United States. After commenting on the history of integrating the armed forces and lauding recent advances, the article concluded: "But there is a problem: the civilian world now lags far behind the military. Said an army brigadier general, 'What worries me is that a military career for a negro is about the top he can get.' A Negro G.I. said it in a different way: the Negro 'begins to see the fellows getting along in the army and begins to say to himself, it would be so goddam nice if it could be like that all over.' "

American soldiers, airmen, sailors, and marines returned from Korea to installations in the United States where they shared barracks, dining halls, snack bars, recreational facilities, and most other aspects of on- and off-duty life. Unlike previous postwar periods when advances won by African Americans during combat quickly evaporated with the return of peace, equal rights for those in uniform actually increased after the cease-fire in Korea.

During the decade that followed the war, blacks made more gains toward equality in the military and influenced civilian progress as well. More than 1.8 million African Americans served in the military during the ten-year period 1946–56 and returned to their homes across American accustomed to the integrated military and expecting to experience the same rights after their discharges. But both blacks and whites quickly recognized that change influenced by the necessities and hardships of combat did not take place as quickly or as easily in time of peace.

The military's last all-black unit disbanded in 1954, and the services, with the exception of the navy, which lagged somewhat behind, recruited African Americans for all specialties. Acceptance in the ranks did not, however, mean acceptance in communities adjacent to military installations. While black service personnel had equal access to integrated military family quarters on bases, they faced the same discrimination in housing in local civilian communities that had always existed.

Black military personnel frequently found off-post housing in neighboring towns restricted to whites only. If available at all, quarters for black service families were dilapidated, located in the poorest neighborhoods, and prohibitively expensive. Colin Powell, later the first black chairman of the Joint Chiefs of Staff, recalled in his autobiography that in 1962 he could find no suitable housing outside the gates of Fort Bragg, North Carolina, and had to send his pregnant wife, Alma, home to live with her parents in Alabama while he attended an army school.

Upon returning from a tour in Vietnam and reporting to the infantry school at Fort Benning, Georgia, in early 1964, Powell, then a senior captain, found the local housing situation for African Americans no better than in North Carolina. Powell recalled in his autobiography, "On my arrival, I immediately set out to find a place for my family to live. I was entitled to government housing when the career course began in the summer. But until then, I needed to find something off-post if Alma and the baby were to join me. Fort Bragg all over again. Plenty of housing available for white officers in the Columbus area. But I was limited to black neighborhoods."

Powell finally found quarters in Phoenix City, across the river in Alabama. Of his tour at Fort Benning, he concluded, "I regarded military installations in the South as healthy cells in an otherwise sick body."

The off-base housing problem was not limited to the South. In Yuma, Arizona, so little housing and other facilities were available for black airmen off base that the air force transferred many to other installations. Calvin A. H. Waller, who reached the rank of lieutenant general and second in command of Operation Desert Storm in 1990–91, discovered after reporting to Fort Lewis, Washington, as a lieutenant in the mid-1950s that the only quarters he could rent were in a remote black neighborhood in Tacoma.

Outside the gates of Topsham Air Force Base in Brunswick, Maine, no landlord would rent to a black regardless of rank, including a veteran pilot with eighteen years of service. Only after the local newspaper campaigned against the discrimination did apartment owners offer the officer a place to live. The *Brunswick Record* later received an award from the New England Weekly Newspaper Association for their assistance in gaining housing for the black officer. In an article on May 18, 1961, the newspaper noted, "While the *Record* takes pride in the award, it came with bitter irony during a week when another Negro airman reported he too was turned away from Brunswick living quarters for himself and his family."

Black service personnel found that education for their children posed as big a problem as housing. All schools for children on military installations were desegregated in 1954, but few of them existed. A large percentage of children of military personnel attended civilian schools off base. Although federal reimbursements supported most of these schools and government subsidies paid for part or all of their buildings, the vast majority of these facilities remained segregated. As a result, many black children of military personnel were transported at government expense, often many miles away, to black campuses.

Not until March 1962 did the secretary of health, education, and welfare rule that segregated schools would lose their federal financial assistance if they did not accept black students in the fall of 1963. Just 26 of 242 affected school districts complied, and they only accepted black children who lived on the military bases, about 10 percent of the total, and not those of families living in the civilian community.

Black military personnel also faced discrimination in furthering their own education. Universities near military installations, especially in the South, refused to accept black students.

Outside the gates of their bases, black military personnel found that civilian communities treated them in the same manner as they did their local minority population. Jim Crow laws, again mostly in the South but to some degree throughout the country, separated black from white in shopping, eating, housing, transportation, and recreational facilities. Frequently these public areas exhibited Whites Only signs, and the towns had police more than willing to enforce these policies.

Shortly after reporting to Fort Bragg in 1963, black army captain Sylvain Wailes noted the discrimination he encountered at his former and current assignments: "When I was at Fort Belvoir, Virginia, there was no decent place to live unless you went into Washington. Housing is segregated around Fayetteville [North Carolina]. I stay on the base for athletics and movies, sometimes go to Raleigh, an hour's drive, for a stage play."

The decade of discrimination that followed the Korean War occurred during the administrations of three U.S. presidents—Eisenhower, Kennedy, and Johnson. Each approached racial issues in a different manner, with varying degrees of intensity. In some cases they influenced change; in others they merely allowed advances, and setbacks, to occur without their direct participation.

Dwight Eisenhower personally favored integration in the services and in the civilian communities and supported federal intervention to ensure equal rights. When he ran for president in 1952, however, he put politics first and in order to gain a majority of the white vote ignored the issue of integration during his campaign. Once elected, he ensured that the services completed the elimination of segregated units and made a few token appointments of blacks as assistants to cabinet officers.

Eisenhower assumed a rather low-keyed approach to racial issues, yet major advances occurred during his administration. The integration of schools on military installations by the Department of Defense resulted from Eisenhower's leadership, but this advancement was superseded and overwhelmed by the May 17, 1954, Supreme Court decision in *Oliver Brown et al. v. Board of Education of Topeka, Kansas*. The unanimous decision ended legally segregated education in the country's public schools.

Some political leaders, anxious to perpetuate the status quo, attempted to define the Court's decision in various ways to maintain a degree of segregation. On May 31, 1955, the Supreme Court restated that they indeed intended *Brown v. Board of Education* to end school segregation and directed that their ruling be instituted "with all deliberate speed."

For all effective purposes, *Brown v. Board of Education* legally struck down the 1896 *Plessy v. Ferguson* decision, ending the "separate but equal" doctrine that had provided the cornerstone of discrimination and segregation. Adherence to the Supreme Court's decision did not, however, occur immediately throughout the country. Some state legislatures and governors resisted or delayed integration to the point of refusing admittance of black children into previously all white schools.

Individual African Americans did not stand idly by while the courts moved toward more equality but rather took actions, both violent and nonviolent, on their own. On December 1, 1955, a Montgomery, Alabama, seamstress named Rosa Parks refused to give up her seat to a white passenger and move to the rear Jim Crow section of a city bus. Parks's arrest produced a bus boycott by Montgomery's black citizens. The nonviolent boycott, backed by court decisions that ruled that segregated public transportation was illegal, led to open seating on Montgomery buses on December 21, 1956.

Eisenhower took a more direct approach to gain equality among the races when his administration teamed with leading Democrats in the Senate, including Estes Kefauver and Albert Gore, both of Tennessee, and Lyndon Johnson of Texas to pass the first civil rights legislation in the United States since 1875. The Civil Rights Act of 1957 granted the attorney general the right to take legal action against anyone denying a person or a group the right to vote. The act also established a Civil Rights Division within the Justice Department and formed a Civil Rights Commission to conduct a two-year study of racial discrimination.

It is impossible to measure the exact influence on civil rights of the 1.8 million African Americans who served in the integrated military during the decade following the Korean War or the impact of the pockets of integration provided by military installations on states dominated by segregation and discrimination. This leadership by the services in integrating society was certainly commendable, but while blacks in uniform enjoyed a greater degree of equality than their civilian brothers, racial problems within the military were far from over.

Military commanders, regardless of the location of their bases, attempted to maintain good relations with local government and civic leaders. Few commanders protested off-post discrimination in housing and other facilities. In fact, nearly two years before Rosa Parks refused to give up her seat in Montgomery, a black air force officer serving in Alabama had also refused to move to the rear of the bus. Instead of supporting his subordinate, the officer's commander issued him a career-ending letter of reprimand, declaring, "Your open violation of the segregation policy is indicative of poor judgement on your part and reflects unfavorably on your qualifications as a commissioned officer."

Some commanders, under pressures from local communities, did not post black military policemen at their front gates or at other places where they might have to deal with the public. At other locations they removed blacks from parade marching units and from athletic teams that participated in local sports contests. Some commanders even removed blacks from funeral details that provided burial support for local veterans.

Blacks also continued to experience discrimination in their integrated units, and they complained to their congressmen and senators about the uneven justice and punishment as well as fewer and slower promotions. Some white officers and NCOs, who maintained their prejudices, placed blacks on the dirtiest and most difficult details and blocked their opportunities for military schooling that might lead to advancement.

The arrival of the 1960s brought increased impatience in the black military and civilian communities. Protests continued, with sit-ins the dominant form of nonviolent action as blacks and their supporters challenged local Jim Crow laws restricting their access to eating establishments and other public facilities.

When President John F. Kennedy was inaugurated in 1961, one of his first actions was to appoint a special assistant for civil rights and to direct Vice President Lyndon Johnson to supervise a committee to guarantee fair hiring practices by the federal government, which included civilian employees of the Department of Defense.

Kennedy noted the few black servicemen in his inaugural parade and in welcoming ceremonies for foreign dignitaries shortly after he took office. He directed immediate actions to increase their public appearances. Kennedy also wanted to improve their quality of life at their assignments. In June 1961 the Department of Defense directed the services to take steps to encourage communities outside their installations to develop plans to integrate housing and other facilities.

The president, like all successful politicians desiring to be reelected, tried to keep voters happy regardless of their color or military connection. Responding to complaints from all sides and being unhappy himself with the lack of implementation, Kennedy formed the President's Committee on Equal Opportunity in the Armed Forces on June, 24, 1962. Kennedy appointed Washington, D.C., attorney Gerhard A. Gesell to chair the committee, which included five prominent white lawyers and educators and three African Americans representing the black press, the Urban League, and the NAACP.

The president directed the committee to provide him advice on how to "improve the effectiveness of current policies and procedures . . . with regard to equality of treatment and opportunity for persons in the Armed Forces and their dependents in the civilian community."

In reality the Gesell Committee worked not for the president but for Secretary of Defense Robert Strange McNamara. On June 13, 1963, the committee made its initial report, beginning with praise for the military's "significant progress in eliminating discrimination among those serving in defense of the nation."

The remainder of the report outlined problems of racial discrimination both on and off military installations and made recommendations for improvement. The report concluded: "Much remains to be done, especially in eliminating practices that cause inconvenience and embarrassment to servicemen and their families in communities adjoining military bases." To meet these issues, the committee recommended that commanders and community leaders form biracial action groups to study solutions for problems in housing, transportation, education, and other public facilities.

The committee did not overlook problems within the military. Its report noted that whites dominated the officer corps and most blacks served in the enlisted ranks. The committee recommended that each military organization appoint an officer to act as a point of contact to receive complaints, conduct investigations, and make recommendations to commanders to remedy racial issues. To ensure equal opportunity for all, the committee recommended that the services periodically review their standards "for promotion, selection, and assignment to make certain that latent ability is always properly measured and utilized."

On July 26, 1963, Secretary McNamara issued Department of Defense Directive 5120.36, which authorized the establishment of the Office of the Deputy Assistant Secretary of Defense (Civil Rights) and urged military commanders to place "off limits" civilian establishments that discriminated against service members. The directive concluded: "Every military commander has the responsibility to oppose discriminatory practices affecting his men and their dependents and to foster equal opportunity for them, not only in areas under his immediate control but also in nearby communities where they may live or gather in off-duty hours."

In addition to responding to the Gesell Committee recommendations, the Department of Defense also instituted equality measures on its own. As late as 1962 ten southern states excluded African Americans from their National Guard organizations. A few segregated army, air force, and naval reserve units also remained in these states. Through direct orders to cease segregation and threats to withhold federal funds, the federal government convinced all the states to integrate their reserve and National Guard units by 1964. However, several of the southern states, especially Alabama and Mississippi, recruited only the minimum allowable number of blacks in their National Guard units.

Military efforts were not occurring in a vacuum, for racial issues dominated American life in the early 1960s. Both the violent and nonviolent advocated various solutions for ending discrimination. Elijah Muhammad, in a speech to fellow Black Muslims in New York on July 31, 1960, called for the creation of a separate black state either in the United States or in Africa. Some black leaders even called for a revolt to forcibly establish equality.

Nonviolent demonstrations, however, reigned over the civil rights movement of the period, with Atlanta, Georgia, minister Martin Luther King Jr. rising to prominence as its leader and most eloquent spokesman. Although Dr. King's work focused on the civilian community, the equal rights for which he strived extended into the military.

King advocated nonviolent action but emphasized that African Americans must become activists to gain equal rights. In 1963, King was jailed in Birmingham, Alabama, for protesting against "injustice" in the city. Some local clergymen believed that King, an outsider, had no right to meddle in the city's difficulties and published a letter in the local newspaper encouraging local blacks not to join the protest but to rely on the courts and political leaders to solve civil rights issues.

From his cell King responded to the clergymen's letter in a lengthy response on April 16, 1963, that more than any other one document sums up the racial problems of the times, what should be done about them, and the reasons behind the black protest movement. According to King, African Americans fell into one of three groups—violent separatists, nonviolent activists like himself, and the others who were too complacent or too downtrodden to take a stand. King classified whites in a similar manner—as violent racists, activists for civil rights, and apathetic or nonactivists.

Early in his letter King provided a moving explanation of the reasons behind the protest movement, noting that African Americans had been waiting for more than three centuries for equal rights. He then discussed the unfairness of segregation and the mistreatment of blacks by whites. King concluded: "There comes a time when the cup of endurance runs over, and men are no longer willing to be plunged into the abyss of despair."

On August 28, 1963, more than a quarter million African Americans and their white supporters held a march on Washington to unite behind civil rights. King delivered a speech describing his "dream" of racial equality, and all across the U.S. Americans began to realize that they would at last have to confront and solve racial issues.

Kennedy, reacting to the march and to his advisers, proceeded as quickly as politically expedient to secure a broad, far-reaching civil rights act, but before its completion an assassin's bullets struck him down on the streets of Dallas, Texas. Lyndon Johnson assumed the presidency, and the civil rights problems and a growing war in Vietnam would soon overwhelm all other issues in the United States and provide the catalyst for more than a decade of violence on American streets and in the rice paddies and jungles of Southeast Asia.

16

VIETNAM: "THE ONLY WAR WE HAD"

From its beginning, the Vietnam War was the most integrated conflict in American history. During the decade following the Korean War, blacks had enlisted in all the services, making the military the most integrated institution in American society. Black soldiers, airmen, sailors, and marines served in Vietnam in every specialty and in every rank.

The United States first became involved in Vietnam during World War II when it supported Ho Chi Minh and the Vietminh rebels in their fight against the Japanese, who were occupying that country. After the war, a succession of U.S. presidents backed first the French and then the South Vietnamese against Ho and his Communist followers, who wanted to take over the entire country: Eisenhower backed the anti-Ho elements because he believed in the "domino theory," that if one country fell to communism, those around it would follow; Kennedy did likewise because he sensed a political need to be "tough on communism"; and Johnson supported the conflict when he took office because, as he later admitted, he did not want to be the first U.S. president to lose a war.

U.S. military advisers arrived in Saigon to assist the South Vietnamese in January 1955. Little American involvement occurred until North Vietnam opened the Ho Chi Minh Trail through neighboring Laos and Cambodia in 1959 to supply their Vietcong army in the south. Later that year U.S military advisers suffered their first casualties to the Communist forces, officially beginning what would become America's longest war.

The United States became even more involved in Vietnam when it supported the overthrow of the Ngo Dinh Diem government on November 1, 1963, in Saigon. After North Vietnamese patrol boats allegedly fired on U.S. navy ships in the Gulf of Tonkin in August 1964, President Johnson retaliated with air strikes against targets in the north. On August 7, the U.S. Congress authorized Johnson to "take all necessary measures to repel any armed attack against forces of the United States . . . and to assist any member or protocol state . . ." of applicable treaties.

The Gulf of Tonkin Resolution passed the House of Representatives 416–0 and the Senate 88–2. America entered the war united to protect South Vietnam from the Communist threat. By the mid-nineteen-sixties the number of U.S. military personnel in Vietnam rose to 60,000 and continued to escalate to a peak of more than 543,000 by 1969.

As they had in every conflict since the Revolutionary War, African Americans saw the military as a place to better themselves while serving their country. In the August 23, 1968, issue of *Ebony* magazine, black psychiatrist Alvin F. Poussaint, quoted in an article entitled "Why Negroes Reenlist," explained the black soldiers' voluntary association with the military:

> Black men in general, particularly from the low socioeconomic groups and particularly from the South, suffer from a low self-esteem because of racism in American society. The black male has always been castrated by the society and has always struggled for a sense of manhood and identity in a white world. Because of the limited opportunities that a racist society allows the black man for achieving manhood, I think many young black men gravitate to the army to prove they are men by risking their lives in combat. Superior prowess in combat is one of the most primitive ways of achieving a sense of manhood. The black man in combat is ready to trade his life for psychological manhood, status, and self-esteem.

Not everyone shared Poussaint's ideas on the military providing the chance for blacks to prove their manhood. Some scholars noted that the armed forces simply offered more opportunity and advancement than any other business, industry, or institution in America.

Whatever their reasons for enlisting, blacks volunteered for the service and then reenlisted, bringing their numbers in the military to a total very near their proportion of the overall population. The Vietnam conflict was the first war in U.S. history in which blacks did not have to fight for the right to fight. From its opening shots, blacks participated in every capacity. African Americans welcomed the opportunity to achieve battlefront equality, and many white Americans, mostly indifferent and certainly unenthusiastic about the war to begin with, were more than happy to have blacks do their fighting.

Early press reports from Vietnam gave high praise to the fighting abilities of black soldiers and the success it reflected on the military's integration policies. *Newsweek*, in an August 22, 1966, article entitled "Great Society—In Uniform," highlighted the outstanding performance of blacks in Vietnam. Hanson Baldwin, military editor for the *New York Times*, wrote on November 20, 1966, that "the Negro has never had it so good in the army."

In its May 26, 1967, issue, *Time* commented on the unity of black and white soldiers in Vietnam and concluded: "For the first time in the nation's military history, its Negro fighting men are fully integrated in combat, fruitfully employed in positions of leadership, and fiercely proud of their performance. More than anything else, the performance of the Negro G.I. under fire reaffirms the success—and diversity—of the American experiment."

On April 24, 1967, Gen. William C. Westmoreland, commander of U.S. forces in Vietnam, addressed the legislature in his native state of South Carolina, which he noted later in his autobiography. "I saw not a black face except those of the janitors standing in the rear. . . ." Nevertheless, Westmoreland told the gathering that the black soldier was serving in Vietnam "with distinction equal to his white comrades in arms."

While the military had ignored or downplayed the valor of black soldiers in the previous wars of the twentieth century, it did not do so now. African Americans were among the first to receive Medals of Honor in Vietnam. On October 22, 1965, Chicago native Pfc. Milton Olive, of Company B, Second Battalion, 503rd Infantry, 173rd Airborne Brigade, earned his country's highest award for bravery by sacrificing his life when he fell on a grenade to save the lives of his comrades in the midst of a battle.

A M-60 machine gun team of the Third Marine Division engages the enemy in Vietnam near the DMZ in September 1966. (U.S. NAVAL INSTITUTE)

Specialist Sixth Class Lawrence Joel, a New York City–born medic assigned to Headquarters Company, First Battalion, 503rd Infantry, 173rd Airborne Brigade, earned his Medal of Honor for repeatedly exposing himself to enemy fire to treat his unit's wounded. Struck several times by bullets and shrapnel, Joel bandaged his own wounds and continued to dispense aid during a fight that lasted for more than twenty-four hours.

Olive and Joel were only the first of many blacks who exhibited great bravery and dedication for the rest of the long war. Of 237 Medals of Honor awarded during the Vietnam War, blacks received twenty—fifteen of them soldiers; five, marines.

Meanwhile, President Johnson attempted to keep all political factions happy during the late 1960s with his Great Society program, designed to promote civil rights and eliminate poverty. The president, genuinely dedicated to advancing the quality of life of all Americans, preferred to focus on domestic issues rather than foreign policy, includ-

Members of the First Cavalry Division exit a UH-1D helicopter during Operation Oregon search-and-destroy mission in Vietnam on April 24, 1967. (U.S. ARMY MILITARY HISTORY INSTITUTE)

ing the war in Southeast Asia. As a result, Johnson increased the number of troops committed to the war zone while never actually articulating a national objective in Vietnam or ever establishing any strategy to accomplish the goal—whatever it was.

To ease political pressures, Johnson at first tried to fight the war only with existing regular forces. When the conflict escalated beyond the regular forces' capabilities, the president authorized an extensive draft to add units to the Regular Army rather than activate the reserves and National Guard.

For a period of several years, Johnson succeeded in fighting the war in Vietnam while at the same time advancing civil rights at home. The Civil Rights Act of 1964, initiated by Kennedy and pushed through Congress by Johnson, provided comprehensive and far-reaching advances in equality in housing and public accommodations. Johnson's Voting Rights Act of 1965 provided blacks more access to the polling places and established regulations to end local and state restrictions designed to prevent blacks from casting their ballots.

As long as the Regular Army and draftees assumed most of the burden of fighting in Vietnam, white America remained mostly detached. The sons of the upper and middle class easily attained school deferments from the draft to attend college and stay on for postgraduate study until reaching age twenty-six, then no longer eligible for the draft. Professional athletes and the sons of the influential secured slots in local reserve and National Guard units, where they endured a few hours of monthly drill at a local armory instead of daily fighting in the paddies and jungles of Vietnam. Some who could not remain in school or gain a coveted slot in the local National Guard unit found doctors, anxious to meet the perceived needs of their communities, to write histories for "football knees," asthma, or other disabilities that qualified for medical deferments.

While white America, especially those from the upper classes, avoided the inconvenience, dangers, and hardships of military life, young black men eagerly volunteered or readily accepted the draft. Albeit low, military pay was better than unemployment, and, as always, the military offered the best opportunity for blacks to advance themselves. Shortly after receiving his Medal of Honor, Lawrence Joel stated his appreciation of the army and the opportunities it provided, saying that black civilians "couldn't make it really big."

However, not all African Americans shared this view of the military in Southeast Asia. From the beginning of the war some black leaders recognized Vietnam as a drain on the energy and budgetary funds needed to advance civil rights. As the war progressed, blacks as a whole realized, too, that they were paying a higher price in casualties than were whites. As one black soldier remarked about the military in Vietnam, "It's the kind of integration that could kill you."

Black leaders advocating violent measures to achieve equality were the first to oppose the Vietnam War. Only later did nonviolent activists join the antiwar movement. One of the initial spokesmen was Malcolm X. In a talk with black teenagers in McComb, Mississippi, on December 31, 1964, Malcolm declared that the United States was "the most hypocritical government since the world began." He explained that the United States was "supposed to be a democracy, supposed to be for freedom and all of that kind of stuff when they want to draft you and put you in the army and send you to Saigon to fight for them—and then you've got to turn around and all night long discuss how you're going to just get a right to register and vote without being murdered."

When the Black Panther Party issued its ten-point program outlining its objectives in October 1966, it included provisions for blacks to fight for their rights at home rather than overseas. Point No. 6 declared, "We want all black men to be exempt from military service."

By the end of 1966 black nonviolent groups, including the Congress of Racial Equality (CORE), the Student Non-Violent Coordinating Committee (SNCC), and Martin Luther King's Southern Christian Leadership Conference (SCLC) joined the antiwar movement. On April 4, 1967, Dr. King denounced the U.S. policy in Vietnam during a speech at New York's Riverside Church and urged young men not to report for military service to the American government, which he declared was "the greatest purveyor of violence in the world today."

King also expressed sentiments he shared with other black leaders in numerous speeches and writings:

> We are taking the young black men who have been crippled by our society and sending them 8,000 miles away to guarantee the liberties in Southeast Asia which they have not found in Southwest Georgia and East Harlem. So we have been repeatedly faced with the cruel irony of watching Negro and white boys on TV screens as they kill and die together for a nation that has been unable to seat them together in the same school. So we can watch them in brutal solidarity burning the huts of a poor village, but we realize that they could never live on the same block in Detroit.

Despite the urging of both violent and nonviolent leaders to refuse induction, the vast majority of blacks reported after receiving their draft notices or volunteered on their own. In early 1967 world heavyweight champion Muhammad Ali publicly refused to answer his draft call. Unlike many other athletes, black or white, who sought medical exemptions or signed up for National Guard units, Ali, claiming to be a conscientious objector, simply refused to report. "No," he said, "I am not going ten thousand miles to help murder and kill and burn other people simply to help continue the domination of the white slave masters over the dark people of the world."

Ali, who later added that "no Vietcong never called me no nigger," had his championship title revoked by the World Boxing Association in April 1967, and in June a court sentenced him to five years in prison and fined him ten thousand dollars. The Supreme Court later overturned his conviction, and Ali remained a hero to blacks and whites in the antiwar movement despite the fact that for every person who refused to report or secured a deferment, some other young man had to take his place.

By mid-1967 the Vietnam War had overtaken the civil rights movement as the leading news story of the year, indeed of the decade. Many whites who opposed the war on moral grounds supported civil rights for the same reasons. Blacks, of course, favored equality and appreciated that Vietnam represented the first fully integrated war in history but nevertheless resented the fact that the conflict detracted from the civil rights movement.

In many ways the antiwar and civil rights movements coexisted without common objectives except a shared antigovernment sentiment and a demand for social change. The rallying cries of both groups came from the civil rights movement. From the riots in the Watts section of Los Angeles in 1964 came the slogan Burn, Baby, Burn, while Stokely Carmichael, head of SNCC, speaking at an antiwar rally in front of the United Nations building in New York in 1967, shouted the words "Hell no, we won't go," which became the anthem of draft evaders of all races. The phrase "black power" apparently originated in Mississippi during a racial confrontation in 1964, but it was Carmichael's repeating of the words in rallies across the country that popularized it among militant and not-so-militant African Americans.

Blacks also were aware that they were suffering more than their fair share of death and carnage in Southeast Asia. During the first years of U.S. combat units' involvement in Vietnam, African Americans, 13.5 percent of the military-age population, made up 10.6 percent of the total force in the war zone. Blacks in general, however, initially served in disproportionate numbers in infantry units that sustained the vast majority of casualties. As a result, in 1965–67, 20 percent of U.S. battlefield casualties in Vietnam were African Americans.

Reasons for this high percentage of black casualties were offered—some rational, others racist. One major factor came from President Johnson's decision to fight the war with the regular forces and draftees

Protesters rally alongside the USS *Constellation* in San Diego in November, 1972. (U.S. NAVAL INSTITUTE)

rather than to call up the National Guard and reserves. Johnson's limited war did not include a commitment from the entire population, and the draft-deferment policies allowed the sons of the wealthy and influential to avoid military service. Twenty years after the conflict, an officer in the cabinet of President Bill Clinton—a president who sat out his own draft eligibility with a student deferment—declared in 1996 that "the best and brightest did not go to Vietnam."

Draft boards faced pools of selective-service registrants who were financially or scholastically unable to get a college deferment, secure "weekend warrior" slots in their local National Guard units, or produce medical waivers. Generally the young men in such straits were black. Some boards, especially the white-dominated ones in the South, experienced no moral dilemma in filling their quotas with disadvantaged African Americans.

In the case of draft boards themselves there was no balancing factor. All across the country blacks were underrepresented on them.

In 1967, of the total of 17,123 draft-board members in the United States, only 261, or 1.5 percent, were black. Not a single African American served on a draft board in the states of Alabama, Arkansas, Louisiana, or Mississippi.

The unchecked draft boards, restricted by the liberal deferment policies and the natural prejudices of many, drafted 30.2 percent of eligible blacks in 1964 compared to 18.8 percent of eligible whites. As the commitment in Vietnam escalated, the percentage of eligible blacks drafted in 1967 rose to 64 percent, compared to 31 percent of whites.

In the early years of the conflict, both blacks and whites were drafted into the army. Not until the late 1960s were any routed to the Marine Corps; the air force and navy relied on volunteers to meet their induction needs. Hampered by inadequate educational opportunities, black draftees generally scored lower on their entrance tests, so instead of assignments to army or Marine Corps technical specialties, which served mostly in the safer rear areas, they ended up in the infantry.

Another factor in the higher causality rate for African Americans was the age-old tradition of blacks volunteering for dangerous duties to prove themselves. Once the services integrated and opened all units to anyone, African Americans volunteered for the combat elites, including the army airborne and the marines. Although no one, regardless of race, wanted to become a casualty, many of the combat units arriving in Vietnam had a disproportionate number of blacks for the simple reason that they had volunteered to be there.

Blacks also reenlisted at a higher rate than whites—in 1965, at a rate of 45.7 percent, compared to only 17.1 percent of whites. These reenlistments, combined with black volunteerism and an inequitable draft, exposed a greater number of African Americans to direct combat. More than two centuries of fighting for the right to fight had finally provided opportunities and equality that soon resulted in black casualties far beyond their "fair share."

Still another reason for the disproportionate number of blacks in the infantry and their resultant casualty rate came from a source that touted opportunities for advancement. In the spring of 1966 projections for rising manpower requirements to meet the needs of the forces in Vietnam showed that the military would need to induct forty thousand draftees every month starting in October. To meet these requirements the Selective Service System had to make choices: end many of the educational deferments, activate National Guardsmen, or lower standards and draft those previously rejected for mental or physical reasons.

Apparently the choice was an easy one for Secretary of Defense Robert Strange McNamara. When faced with the politically sensitive issue of ending deferments, held mostly by middle- and upper-class whites, or adjusting standards to make more lower-class whites and blacks eligible for the draft, McNamara devised a plan to make it appear he was helping the lower classes improve themselves.

McNamara announced in a speech in New York City that his new Project 100,000 (POHT) would enlist men who had previously been declared ineligible because of low mental test scores. McNamara proclaimed that the project would provide for the "subterranean poor" and end the "idleness, ignorance, and apathy" of their lives. According to the secretary, these young men "have not had the opportunity to earn their fair share of this nation's abundance, but they can be given an opportunity to return to civilian life with skills and aptitudes."

McNamara made no mention of African Americans in his announcement, but in a later speech he described POHT as a means of restoring self-respect for blacks. He explained: "What these men badly need is a sense of personal achievement—a sense of succeeding at some task—a sense of their own intrinsic potential. . . . They have grown up in an atmosphere of drift and discouragement. It is not simply the sometimes squalid ghettos of their external environment that has debilitated them—but an internal and more destructive ghetto of personal disillusionment and despair: a ghetto of the human spirit."

According to McNamara, POHT's New Standards Men (NSM) would train in the technological and military specialties. They would acquire skills, self-discipline, and veterans' benefits to help them make a contribution to their communities after their discharge.

The Department of Defense informed the services that each would accept POHT draftees or volunteers. The army was assigned a minimum quota of 25 percent of the NSMs, the marines, 18 percent; and the air force and navy, 15 percent each. Between October 1966 and June 1969, POHT accepted 246,000 previously rejected recruits. More than 80 percent were high school dropouts, 40 percent read below the sixth-grade level, and 15 percent functioned below the fourth-grade level. More than half scored less than 85 on their IQ tests. Slightly more than 40 percent of all POHT enlistees were African Americans.

From its beginnings POHT produced more problems than solutions for both the recruits and the services. The NSMs required extended training time at additional cost, and few ever mastered anything beyond basic skills. Most of the NSMs had to recycle through training, many never satisfactorily completed their courses at all. Only 68 percent advanced from basic to advanced training. Fellow recruits, as well as some of the training cadre, referred to the NSMs as "the moron corps."

Black and white POHT recruits experienced about the same difficulties and failures during and after training. More than a third, some 120,000, received discharges before completion of their enlistment for unauthorized absences or other disciplinary reasons. Of these, 80,000 received less than honorable discharges—far short of the skills training and discipline that McNamara promised they would have when they returned to their civilian communities.

More than 37 percent of the NSMs ended up as infantrymen in Vietnam. Nearly half of this number were black, which ultimately increased the casualty figures for African Americans.

McNamara, leaving office in 1968, did not stay in the secretary's position to see either POHT or the Vietnam War to a conclusion. He later admitted that he never had any confidence that the United States could win the war. Perhaps that is the reason he permitted the wholesale deferment of the best and brightest and instituted a program to place the NSM in uniform to fight American's most unpopular war.

POHT quotas dropped to seventy-five thousand in 1970 and fifty thousand in 1971, before the program's termination, when the United States withdrew the majority of its troops from Vietnam in 1972. By that time 360,000 NSMs had been enlisted through POHT, about 145,000 of whom were black.

Despite the disproportionate 20 percent casualty figures that blacks absorbed in the first years of the war, their performance remained loyal and diligent. The Vietcong and North Vietnamese made propaganda efforts to get black soldiers and marines to lay down their arms—to no avail.

Blacks and whites in the integrated units shared the hardships and dangers of the battlefield and developed bonds that only those who share combat ever form. Soldiers and marines in the field spoke of the only color being the olive drab that they all wore. In his biography General Westmoreland claimed: "I am convinced the United States

A marine of the Twenty-sixth Marine Regiment lights the cigarette of a soldier from the Seventh Cavalry during joint Operation Pegasus in Vietnam in April 1968. (U.S. ARMY MILITARY HISTORY INSTITUTE)

never fielded a more professional force than in South Vietnam during the years 1966–69."

Although cooperation existed between blacks and whites in combat, unrest at home over civil rights and the antiwar movement began to affect the morale of those in Vietnam. The 20 percent casualty figure became a rallying point for those who objected to blacks shouldering more than their share of the war's burden.

The military also recognized the disparity and in 1967 began to shift more blacks to support positions. Completion of tours also decreased the number of African Americans in the air cavalry, airborne, and marine divisions that had arrived in Vietnam with a high percentage of minority soldiers. By the end of the year, black casualties decreased to 13 percent and grew smaller during each remaining year of the war. At the conclusion of the conflict, a total of 58,151 Americans had been killed in Vietnam, 7,115 of whom were black.

The 12.2 percent equaled the proportion of blacks in the total U.S. population.

All in all, the military showed a remarkable sensitivity to black complaints about casualties and took direct action to redress the imbalance. Activists who opposed the war and those who resented the finances and attention the conflict took away from the civil rights movement nevertheless continued to quote the 20 percent casualty figure as if it remained true for the duration of the war. Some current publications still misquote the statistics.

As late as 1967 black servicemen reenlisted at a rate double that of whites. They volunteered to remain in the service past their first tour of duty because of the opportunities and equality that existed in uniform. By the end of 1967, however, reports of disproportionate casualties, combined with black leaders' opposition to the war, began to take its toll on the morale of African Americans in the military.

The major U.S. media, black and white, noted the lack of support for the war by minority leaders. Thomas A. Johnson, a black reporter for the *New York Times*, wrote, "This is the first time in the history of America that national Negro figures are not urging black youths to take up arms in support of American policy to improve the lot of the black man in the United States."

In Vietnam black and white Americans fought the war to the best of their ability, often offering the sarcastic explanation "It is the only war we have." Whites and blacks at times responded to reporters' questions as to why they got along in the field by shrugging their shoulders and commenting, "Out here, no matter what our color, we're all niggers."

The racial dissent at home, however, reached the war zone. Most significant was the assassination of Martin Luther King Jr. in Memphis on April 4, 1968. Rioting broke out in more than one hundred American cities, and federal troops and National Guardsmen were called out to quell the racial unrest. Some of King's followers continued to promote his nonviolent philosophy, but as a whole, African Americans were no longer willing to wait for equality; they now demanded it and backed up those demands with violence. This tension in the streets of America soon spilled over into the U.S. military forces in Southeast Asia.

In several rear support areas in Vietnam blacks gathered to honor King and to protest racism. Some whites reacted by displaying Con-

A soldier in the 199th Light Infantry Brigade in Vietnam reads a letter from home in February 1970. (U.S. ARMY MILITARY HISTORY INSTITUTE)

federate flags and in at least one case burned a cross in front of a barracks that primarily housed blacks. A few weeks after King's assassination, black soldiers, who made up about half of the prisoners, rioted at the army's confinement facility at Long Binh.

After King's death, Vietnam was no longer an island of racial cooperation. The violence that dominated Americans in the United States and at military installations abroad polarized the races in uniform more than ever before. Many white commanders charged that blacks were troublemakers, and some black soldiers, as well as civilians, referred to Vietnam as the war where "black men killed yellow men on orders from white men who represented a land they had taken from red men."

Other problems at home also spilled over into Vietnam. The proliferation and acceptance of drugs in the civilian society became a military problem, black and white, as well.

More important, however, was the antiwar movement that demanded that the United States withdraw from Southeast Asia. The morale of those in the war zone was damaged by students displaying Vietcong flags at protest rallies and business and political leaders declaring the war immoral and its participants murderers. Veterans, black and white, returned home not to bands and parades but to protesters calling them capitalist pigs and baby killers.

Most African Americans in Vietnam appreciated that the military was more integrated than the rest of the country, but they also were well aware that their infractions received punishment more often and more severely than did those of their white comrades. They also faced discrimination in promotions and awards. Some white commanders and senior NCOs still held the racist attitudes of their upbringing and misinterpreted the positive "black pride" of their men as if it were troublemaking "black power."

By 1970 the majority of the American public favored withdrawal from Vietnam. Most of the antiwar protesters extended their contempt for the conflict to those who fought it. For the first time in American history the media and the public vilified military personnel regardless of color. Responding to political demands, Richard Nixon, the newly elected president, began pulling the military out of Vietnam in 1969, and by 1970 the trickle increased to a flood as the United States turned over the responsibility for the war to the South Vietnamese.

While polarization and racial confrontation had occurred among support troops on a limited scale behind the lines, where boredom presented more danger than the enemy, combat troops in the field, now despised by the very people at home who had sent them to war, generally remained united to fight the Vietcong and North Vietnamese. As the conflict drew to an end, soldiers and marines worked together to ensure their mutual preservation. No one, black or white, wanted to be the last American killed in Vietnam.

While racial confrontations and divisiveness did occur in Vietnam during the final years, combat bonding, particularly among field units, limited large-scale racial incidents. At bases outside the war zone, including those in the United States, commanders had to institute special security units and curfews to limit fighting between off-duty blacks and whites. Racial epithets frequently triggered confrontations that ranged from fights between two people to large riots involving hundreds.

During the first eight months of 1968, marines at Camp Lejeune, North Carolina, reported 160 incidents of assaults, muggings, and robbery with racial overtones. At Goose Bay, Labrador, in March 1971 blacks and whites fought over the music selection at their service club, and a similar disagreement led to four days of rioting at Travis Air Force Base, California, the following May.

More than two hundred significant racial incidents occurred in the army in 1970, with eighteen large confrontations at U.S. installations in Germany. On July 4, 1970, in a "call for justice," more than a thousand black soldiers assembled on the campus of Heidelberg University to protest discrimination in military courts-martial, assignments, promotions, housing, and recreational facilities.

Despite the divisiveness of the Vietnam War among the American people and within the services themselves, the conflict, along with the concurrent civil rights movement, did force advancements and improvements for blacks in the military. The need for additional personnel to meet the wartime requirements, combined with the relative ease with which many whites of draft age secured medical and educational deferments, provided blacks their best opportunity in history to join the armed services. Despite lingering discrimination, blacks could serve in any position and advance through the ranks in accordance with their ability. Although reenlistments of blacks and whites steadily decreased as the war extended into the 1970s, blacks remained twice as likely as whites to remain in the service past their first enlistment.

During Vietnam one of the major complaints by black enlisted personnel was the lack of officers of their race. Improvements made during the war in increasing the numbers of African-American officers and their responsibilities produced junior leaders who have gone on to become today's senior military commanders.

Artillerymen of the Third Marine Division prepare their 105 mm howitzer for a fire mission in Vietnam in September 1968. (U.S. NAVAL INSTITUTE)

Adding more black officers was an issue that could easily be solved. In 1962, African Americans made up only 1.6 percent of the officer corps—3.2 percent of the army, 1.2 percent of the air force, and less than 1 percent of the navy and Marine Corps. Most black officers in all the services were lieutenants or captains. Air Force major general Benjamin O. Davis, near the end of a long and illustrious career, held the only flag rank by a black in the entire military. The army had only six black colonels, and the navy had but three full commanders. Seven captains represented the highest-ranking blacks in the Marine Corps.

Six years later, in 1968, blacks made up 9.7 percent of the total enlisted personnel—11.4 percent of the army, 4.8 percent of the navy, 11.3 percent of the Marine Corps, and 10.3 percent of the air force.

These percentages would remain fairly stable for the rest of the war. Black officers increased from 1.6 percent in 1962 to 2.1 percent in 1968—3.4 percent of the army, 0.4 percent of the navy, 0.8 percent of the marines, and 1.8 percent of the air force.

The number of black officers increased, albeit slowly, each year of the war, reaching 2.3 percent in 1971. Their added responsibilities also led to advancement in rank. By 1973 the army had twelve black generals, the air force touted three, and the navy had its first African-American admiral.

The number of African Americans attending the military academies also increased during the Vietnam War. Some of the progress resulted from the academies' attempts to provide equality as they increased the numbers in classes to meet the rising needs of the services. Noteworthy is that opposition to the war in Vietnam and the general public contempt for officers and all things military produced the lowest number of applicants to the academies in their modern histories during the late 1960s and early 1970s.

The U.S. Military Academy averaged less than four black graduates per year from 1963 through 1967. In 1968, West Point established an Equal Admissions Opportunity Program. The objective was to increase the number of minority students and ensure that the corps of cadets was representative of the national population. For the first time, the academy actively recruited African Americans. By 1969 black cadets numbered forty-five, as compared to only nine the year before. Twelve blacks graduated from West Point in 1972 and thirty-six in the class of 1976.

The U.S. Naval Academy also increased its admissions of blacks, and by 1971 the number of African-American midshipmen totaled fifty. Annapolis commissioned twelve black officers in 1972, and this number increased to thirty-nine in 1976. In 1968 the navy also established its first ROTC unit at the all-black college at Prairie View A&M in Texas. Two years later the school commissioned its first thirteen naval officers.

Congress authorized the establishment of the U.S. Air Force Academy at Colorado Springs, Colorado, on April 1, 1954, and the school accepted its first cadets in the summer of 1955. Four years later three African Americans entered the academy and graduated with the class of 1963. The numbers of black cadets continued to rise, with thirty-one earning their commissions in the class of 1976 and new classes averaging twenty-four African Americans each.

Black women also benefited from the additional numbers required because of the Vietnam War and the reluctance of middle- and upper-class whites to enlist. In 1967 the Department of Defense lifted the 2 percent ceiling on the number of women in each of the services established shortly after the end of the Korean War. It also opened additional specialties for females and placed no quotas or additional restrictions on black women. By 1971 black women officers numbered 431, or 3.3 percent of all female officers in the armed forces. Black enlisted women totaled 4,236, or 14.4 percent of the services' female strength.

America's first completely integrated war was also its most unpopular. The Vietnam War divided the United States more than any single event since the Civil War; its influence did not fade with the conflict's conclusion. The war played an important role in the civil rights movement, and the civil rights movement played an important role in the war. Blacks, as in every American conflict, played a critical role in the conflict, and their sacrifices were significant—all the more so considering the times and the attitudes at home.

Wallace Terry, in one of the few books about African-American participation in the Vietnam War, wrote of the black in the conflict:

> He fought at a time when his sisters and brothers were fighting and dying at home for equal rights and greater opportunities, for a color-blind nation promised to him in the Constitution he swore to defend. He fought at a time when some of his leaders chastised him for waging war against a people of color, and when his Communist foe appealed to him to take up arms instead against the forces of racism in America. The loyalty of the black Vietnam veteran stood a greater test on the battleground than did the loyalty of any other American soldier in Vietnam; his patriotism begs a special salute at home.

17

AFTER VIETNAM: UNREST AND THE ADVANCE OF EQUALITY

The conclusion of the conflict in Southeast Asia did not bring any real peace to the U.S. military. American forces still faced Soviet and other Communist foes in Europe and Korea, and the threat of nuclear war inflamed military expenditures and planning. Compounding the international troubles was the demoralized, polarized, and drug-infested condition of the U.S. military. Many officers and senior enlisted men retired or resigned rather than meet the challenges of rebuilding the services in the postwar era.

For the services to meet the challenge of the Soviet Union's objective to spread communism around the world, the U.S. armed forces had to first resolve their internal problems. Racial tensions dominated these difficulties, and for the first time when a war ended, the U.S. military focused on advancing relations between whites and blacks instead of ignoring the issue. Leaders finally accepted that the blacks' "place" in the military was in any position and at any rank for which they qualified. Military commanders also understood that they had the responsibility to ensure that all service personnel received equal treatment and equal opportunity for advancement.

The evidence that corrective measures must take place was obvious, and for once the military willingly admitted that racial problems existed within the ranks. An article in the April 1970 issue of *Army Digest* entitled "Solders Look at Race" noted: "The barrier of color remains the most significant point of friction and misunderstanding in the military today."

In October and November 1972 racial incidents occurred among crews aboard the carriers *Kitty Hawk* and *Constellation* and the oiler *Hassayampa*. Confrontations became so severe on the *Constellation* that the carrier had to return from training at sea to its San Diego port to stabilize the situation. Most of the complaints of the black sailors were similar to those of soldiers in Germany, who had protested that their mostly white officers either did not, or could not, communicate with them to identify and solve problems.

Communication played an important role in the Department of Defense's efforts to establish racial equality and harmony. To ensure that an environment existed to nonviolently solve issues and complaints, the military instituted a series of seminars, studies, and orders. These efforts began prior to the end of the war in Vietnam and often in the midst of direct confrontations between the races during the early 1970s.

In November 1969, General Westmoreland, now chief of staff of the army, ordered that every unit institute a series of seminars on interracial relations. A month later, Secretary of Defense Melvin H. Laird issued Directive 1100.15, "Equal Opportunity Program," which mandated that all Department of Defense components establish an equal-opportunity program and develop affirmative-action plans. The directive authorized commanders to deny access to their installations to any organizations that did not support equal opportunity, including civilian contractors. Secretary Laird also directed the establishment of the Interservice Task Force on Education in Race Relations to develop instruction for all the armed forces.

To support ongoing equal-opportunity efforts and provide additional resources for communications, Secretary Laird issued another order on June 24, 1971. Directive 1322.11, "Education in Race Relations for Armed Forces Personnel," established the Race Relations Education Board (RREB) to develop policy guidance for the program and authorized the formation of the Defense Race Relations Institute (DRRI) to train instructors and produce instructional materials in race relations.

The DRRI staff, headquartered at Patrick Air Force Base, Florida, went through several phases in devising an educational program to promote understanding of ethnic differences and increase racial harmony.

The DRRI staff had little reference material available from published sources or civilian schools to help them accomplish this huge, complex mission. Most of the materials and instructional techniques they developed, much by trial and error, pioneered the way for race-relations education throughout the country, work which schools, business, and industry would eventually adopt.

DRRI graduates, armed with the school's instructional materials, returned to their units to conduct mandatory race-relations training. One of the documents, "Understanding the Minority Member in Uniform Today: The American Black," provided an excellent summary of the post–Vietnam War African American in uniform:

> A black in uniform does not cease to be black. Like any other serviceman, he seeks to retain his personal and racial identity while identifying and relating to the service.
>
> For generations, the black American has perceived himself to be without identity. Personal success was usually achieved by accommodating and imitating whites; therefore, the black was expected to conform and act as if he were invisible and inferior. The younger generation of blacks wishes to be considered equal while being different. Thus, the logical emergence of a thrust for identity. The young black serviceman strongly desires and needs the opportunity to express his views with young whites and with older black and white NCOs and officers. He has seen the injustices in the nation's schools, courts, law enforcement agencies, and in his quest for economic security; consequently, he is socially aware and sensitive to any form of discrimination, real or imagined. Due to the social patterns in the society from which he emerged, he frequently maintains close relations with his black brothers, resulting in voluntary polarization. Some young blacks consider violence an acceptable course of action for resolving conflict.
>
> A communications and credibility gap exists between young blacks and whites and, perhaps even more so, between young blacks and older black and white NCOs and officers. Having experienced social injustice in their communities, young blacks out of contact with whites and supervisors may be influenced by informal group leaders and may perceive inspectors general, chaplains, legal assistance officers, commanders, and NCOs as institutional representatives.

> Promotion and judicial procedures may also be regarded with suspicion. Complaints emanating from such perceptions are considered to be symptoms of deeper problems related to a lack of awareness, understanding, and above all, communications. This average young black serviceman resents having his newly found racial pride confused and interpreted as evidence that he is a black militant or a racist. He realizes that the new problem facing the service is protecting him and his white comrades from the influence of both the black and white racist. This point is extremely important because the increase in aggressiveness and vocalism of the black serviceman is not necessarily a problem. It only becomes a problem when commanders overreact and write off these blacks as troublemakers and militants.

The armed forces reinforced the increased communication on racial issues with direct action. Dining facilities began weekly "soul food" nights and added traditional black dishes to daily menus. In response to complaints of black servicemen and women, the Army and Air Force Exchange Service authorized $1 million in 1970 to train six thousand barbers and eleven hundred beauticians to cut and style the hair of African Americans. Base and post exchanges around the world also added black-oriented products and supplies for service personnel and their family members. In an article in the July 1971 edition of *Ebony*, a writer noted after a visit to Westover Air Force Base, Massachusetts: "Black airmen and their families now find one product after another that has been ordered with them in mind. There are various conditioners and oil sprays for Afro hair styles; facial cosmetics and colognes, dashikis, greeting cards with 'soul' messages and drawings, magazines, phonograph records, and the like."

The Department of Defense did not overlook off-post discrimination. A 1972 policy authorized Housing Referral Offices (HROs) on each major installation to maintain listings of off-post properties for rent and sale. These listings included information on the equality practices of landlords and real estate sales offices. The HROs had the authority to place properties off-limits, to investigate complaints, and to resolve disputes.

New regulations required that each unit have a Race Relations Council to hear complaints, resolve issues, and enhance communication. The chain of command monitored closely each unit's compliance with policies for working toward racial equality. Efficiency reports on officers and NCOs, the basis for promotions and assignments, now included a section on the individual's support of race relations and equal opportunity.

Commanders at all levels also made it clear that racism and discrimination would no longer be tolerated. Adm. Elmo R. Zumwalt Jr., chief of naval operations, issued a memo in December 1970 directing everyone, regardless of rank, "to help seek out and eliminate those demeaning areas of discrimination that plague our minority shipmates." Zumwalt concluded, "Ours must be a navy family that recognizes no artificial barriers of race, color, or religion. There is no black navy, no white navy, just one navy—the United States Navy."

Shortly after becoming the commandant of the Marine Corps, Gen. Robert E. Cushman Jr. issued instructions to his subordinate commanders on June 6, 1972, concerning human-resource training. Cushman declared that the marines were "in the front line of the Nation's effort to improve the areas of understanding and cooperation among all Americans" and emphasized that improved race relations were critical in maintaining the corps' proficiency.

Commanders of the other services shared Cushman's belief that combat readiness depended on racial harmony. They also understood that they could not easily or quickly eliminate racial hostilities because both white and black servicemen and women were reflecting and perpetuating biases learned in their civilian communities, places still lagging far behind the equality practices instituted by the military. While commanders could not be responsible for beliefs and prejudices, they could control acts of discrimination and racism. Furthermore, they could and did work to educate both races and discipline those who refused to cooperate.

For the first time, the senior military and civilian leadership stood united in demanding the end of discrimination and the enforcement of equality for all. Each of the service chiefs issued statements of zero tolerance of racism of any type. Anyone not in compliance would have no place in the military of the future. General Cushman, in a letter to his subordinate commanders on July 31, 1972, expressed the feeling of the military's leadership: "Those individuals who cannot or will not abide by this principle should seek other employment. There is no room for such marines in our corps today."

Prejudices, of course, could not be ended overnight, but information and education could lead to new opinions and attitudes. In a speech on June 21, 1973, General Cushman observed:

> Each new marine we get—whether officer or enlisted—brings along, figuratively speaking, his own personal seabag filled with the prejudices he has been collecting for eighteen years or more. The simple act of putting on a green uniform does not cause him to empty that seabag. But through training we can instill the desire in him to repack that seabag—discarding the harmful preferences and prejudices—so it does both him and fellow marines the most good."

The services adopted the Department of Defense guidelines to increase communication, racial equality, and the overall quality of life of all military personnel. Most of these reforms achieved their objectives.

The combined efforts of increased communication, direct action, and zero tolerance of racism began to yield dividends. A few senior commanders and NCOs who did not agree with changing race relations and equal-opportunity advancement either retired or resigned, but as a whole the military accepted the changes in the spirit of being the right thing to do, if for no other reason than it contributed to combat readiness.

Interestingly, the loudest complaints about the advances in race relations in the military came not from internal sources or civilian organizations but from the U.S. Congress. Louisiana Democrat F. Edward Hebert, chairman of the House Armed Services Committee, accused naval leaders of permissiveness and on November 13, 1972, announced an investigation of "alleged racial and disciplinary problems" in the navy. Although the other services were conducting the same activities as the navy, the high visibility of Admiral Zumwalt gained him more notice from the press, which apparently drew Hebert's attention and wrath.

Hebert, who had actively opposed the Civil Rights Act of 1964, conducted his investigation behind closed doors. However, leaks soon revealed that the congressman was more interested in blaming his perception of a decrease of discipline on the advances of blacks in the navy rather than on discrimination or racism. In fact, Herbert's committee's final report, released on January 2, 1973, made sweeping denials of any racism in the navy and blamed the racial turmoil on an increase in permissiveness and a decrease in discipline.

Most Americans, and especially those in the military, recognized Hebert's investigation and report as a last-gasp effort by a southern congressman to keep blacks in what he perceived as "their place." Apart from serving as an embarrassment to all concerned, the report had no effect.

At the same time Hebert's committee investigated race relations in the military, the Department of Defense was concluding its own study. Its report acknowledged that discrimination resulted from long-established racism in the society as a whole and that it was impossible for military personnel not to be effected by it. The study identified problems in unequal military justice, promotion practices, and assignment polices and made recommendations to improve those areas and move the military closer to true racial equality. The study also recommended that "Americans of Spanish descent" be included in future efforts and studies.

One of the most crucial factors in racial unrest and negative attitudes toward the military during the Vietnam era was the Selective Service System. Blacks believed that boards drafted them in disproportionate numbers and that liberal deferments were more readily available to whites. At the center of the antiwar movement were draft-age whites who opposed their own possible induction as much as, or more than, the conflict itself.

President Richard Nixon began measures to end the draft shortly after his inauguration. On March 27, 1969, he announced the formation of a commission headed by former secretary of defense Thomas S. Gates, with the mission to develop "a comprehensive plan for eliminating conscription and moving toward an all-volunteer armed force."

While the Gates Commission conducted its study, Nixon took more immediate steps to make the draft more fair. On May 13, 1969, Nixon asked Congress to amend the selective service laws to allow the process to convert to a lottery system based on birth dates. The law phased out various waivers, with college deferments ending in December 1971 as the lottery, or "Vietnam Bingo," designated who would serve.

In January 1970 the Gates Commission reported that it had identified several issues that might affect the success of the all-volunteer armed forces. These issues included concerns about a decline in patriotism if the military did not include all Americans, leading to the possibility that the armed forces would be eventually dominated by adventurers and mercenaries. The commission also identified another issue—the belief by some Americans that an all-volunteer force might lead to a military dominated by blacks and enlistees from low-income backgrounds "motivated primarily by monetary rewards rather than patriotism."

In its conclusions, however, the Gates Commission dismissed all of these concerns and declared, "We unanimously believe that the nation's interests will be better served by an all-volunteer force, supported by an effective stand-by draft, than a mixed force of volunteers and conscripts."

As the services reduced their numbers in the post-Vietnam era, President Nixon and Secretary of Defense Melvin Laird prepared to end the draft and institute the all-volunteer force. As to questions about the type of volunteers who would make up the army, Laird, in his report to the president on July 20, 1972, explained, "We do not foresee any significant difference between the racial composition of the All-Volunteer Force and the racial composition of the Nation. . . . We are determined that the All-Volunteer Force shall have a broad appeal to young men and women of all racial, ethnic, and economic backgrounds."

Nixon approved the all-volunteer force, and on June 30, 1973, the last official conscript, following more than 17.5 million Americans who had been drafted since the Civil War, was inducted into the army. Two years later, every serviceman and woman in the U.S. armed forces was a volunteer.

To encourage enlistments, the military increased the pay for all ranks and funded monies to make installations more livable by improving housing, recreational facilities, and retail outlets. Recruiting budgets soared as the military advertised through television, radio, and print media to attract volunteers.

In the same manner that African Americans had come forward in every U.S. war for more than two centuries to volunteer, blacks filled the recruiting stations and enlisted in the all-volunteer force. Attracted by the opportunity to learn new skills, receive fair compensation, and serve their country in the most integrated society in America, blacks viewed the military as a way to enhance their current status and build their future.

During the last full year of the draft in 1972, blacks made up 12.6 percent of the armed forces' enlisted ranks—17 percent of the army, 12.6 percent of the air force, 6.4 percent of the navy, and 13.7 percent of the Marine Corps. These numbers experienced a steady upward growth in the all-volunteer force, with blacks reaching 22.1 percent of the enlisted total force in 1981—33.2 percent of the army, 16.5 percent of the air force, 12.0 percent of the navy, and 22 percent of the Marine Corps.

The number of black officers also increased after the first decade of the all-volunteer force, but not at the same accelerated rate as the enlisted percentage. In 1972, 2.3 percent of the armed forces' officers were black; 3.9 percent in the army, 1.7 percent in the air force, 0.9 percent in the navy, and 1.5 percent in the Marine Corps. By 1981 the number of black officers had increased to 5.3 percent of the armed forces' total: 7.8 percent, army; 4.8 percent, air force; 2.7 percent, navy; and 4.0 percent, Marine Corps.

The National Guard and reserves also significantly increased the number of blacks in the postdraft period. By the end of 1982, African Americans constituted 4.4 percent of reserve-component officers and 19.4 percent of the enlisted force.

Despite the relatively low numbers, black officers were advancing through the ranks. In 1971 the navy promoted Samuel L. Gravely Jr. as its first black rear admiral. Gravely would later advance to vice admiral before retiring. Daniel "Chappie" James became the first black four-star general in the air force in 1975; Roscoe Robinson Jr., the army's first black full general in 1982. By 1984 there were twenty-two active-duty black general officers in the army, eight in the air force, two in the navy, and one in the Marine Corps. In addition, Clifford L. Alexander became the first black secretary of the army in 1977, and Cadet Vincent K. Brooks became the first black to serve as the first captain and brigade commander at West Point in 1979.

The number of black women in the armed forces also increased during the decade following the Vietnam War. Affirmative-action plans within the military and the openings of many occupational specialties increased the number of women from 3.3 percent of the officers and 14.4 percent of the enlisted ranks in 1972 to 10.3 percent of the officers and 27.4 percent of the enlisted personnel in 1981. In 1979 the first black-woman flag officer, Hazel Winifred Johnson, pinned on her star as a brigadier general and became the chief of the Army Nurse Corps.

The rapid increase of African Americans in the military produced mixed reactions from whites and blacks. Many whites, extremely happy to no longer have to worry about the draft, welcomed the all-volunteer concept and cared little about who filled the ranks as long as it was not they or their sons. Some whites expressed concerns about blacks dominating the armed forces and the possibility of reverse discrimination if the numbers of African Americans in uniform continued to increase. Some whites also brought up the old question of whether a black-dominated military would willingly fight an enemy "of color."

The black community also had mixed reactions. Most welcomed the opportunities of military service and its degree of equality. Some, however, noted that, as in the early years of Vietnam, if the United States went to war, African Americans would suffer far more than their proportionate share of casualties. Most black leaders believed that the benefits were worth the risks. Cong. Ronald V. Dellums of California, in an article entitled "Don't Slam Door to Military" in the June 1975 issue of *Focus*, wrote: "Black volunteers understand what joining the military means. If, through experience of free choice by individuals, there are more blacks in the service than in the population, we should expect a proportionately greater sacrifice. The whole idea of a volunteer army is that the individual will take this risk and this responsibility on by his or her free choice."

Increased communication between the races in uniform, direct actions by commanders to advance equality, and the fact that no one served involuntarily reduced racial tensions. Incidents declined as the military reestablished discipline and professionalism.

In July 1979 the military formally expanded its race-communications efforts to include all human relations, and the Defense Race Relations Institute changed its name to the Defense Equal Opportunity Management Institute. Civilian schools and businesses looked to the military as a paradigm for establishing communication and maintaining racial harmony, for the Department of Defense was the country's most integrated and equitable organization.

On May 18, 1981, the Department of Defense reissued a charter entitled "Human Goals" that outlined the military's stance on equal opportunity and affirmative action. Signed by the secretary of defense and each of the service secretaries and military chiefs, the charter first reiterated the American principle that "the individual has infinite dignity and worth."

It then set forth its objectives:

> In all that we do, we must show respect for the serviceman, the servicewoman, and the civilian employees, recognizing their individual needs, aspirations, and capabilities.
>
> The defense of the Nation requires a well-trained force, military and civilian, regular and reserve. To provide such a force, we must increase the attractiveness of a career in Defense so that the service member and the civilian employee will feel the highest pride in themselves and their work, in the uniform and the military profession.
>
> The attainment of these goals requires that we strive:
>
> To attract to the defense service people with ability, dedication, and capacity for growth;
>
> To provide opportunity for everyone, military and civilian, to rise to as high a level of responsibility as possible, dependent only on individual talent and diligence;
>
> To make military and civilian service in the Department of Defense a model of equal opportunity for all regardless of race, color, sex, religion, or national origin and to hold those who do business with the Department to full compliance with the policy of equal employment opportunity;
>
> To help each service member in leaving the service to readjust to civilian life; and
>
> To contribute to the improvement of our society, including its disadvantaged members, by greater utilization of our human and physical resources while maintaining full effectiveness in the performance of our primary duties.

18

UNITED DEFENDERS: TODAY AND TOMORROW

As the military improved race relations and advanced equal opportunity for all those in uniform, the United States as a world power continued to assume those responsibilities necessary to preserve democratic freedoms abroad. "No more Vietnams" became the watchword for the American public and political establishment in the postwar decade, but the United States did react to crises and took stands when necessary. Blacks and whites, all volunteers, now stood united in the most cohesive military in U.S history to defend the country and project American military power around the globe.

The first major use of the all-volunteer military came on October 25, 1983, when a joint U.S. task force, at the request of the Organization of Eastern Caribbean States, invaded the island of Grenada to rescue American medical students and to restore the country's democratic government, which had been replaced by Communists supported by Cuban advisers and arms from the Soviet Union. In only a few days Operation Urgent Fury accomplished all its objectives at the cost of 18 Americans dead and 116 wounded. Grenada lost 45 soldiers killed, with Cuban losses numbering 24 dead, 59 wounded, and 650 captured.

Not everything worked perfectly in Grenada. The services experienced problems in intelligence, communications, and coordination, but most agreed with Secretary of Defense Caspar Weinberger, who declared, "In both military and political terms, the operation on Grenada was a success."

In addition to restoring democracy to Grenada, Operation Urgent Fury restored Americans' faith and confidence in their military—an all-volunteer force composed of more than 30 percent African Americans.

On December 20, 1989, the United States launched its largest military operation since Vietnam—to install the democratically elected government of Panama; to capture the country's dictator, Manuel Noriega; and to neutralize the Panamanian Defense Force (PDF). Once again the totally integrated, all-volunteer U.S. military quickly accomplished its objectives, with the loss of 23 killed and 347 wounded.

The biggest news in respect to both Grenada and Panama regarding black military history was that neither conflict generated press coverage about white and black personnel but rather about American servicemen and women performing their jobs and doing them well. After two centuries of warfare, blacks were full and equal partners in the defense of their country and the world's democracies.

The fall of the Berlin Wall and the collapse of the Soviet Union in the early 1990s left the United States as the only world power, but enemies and responsibilities remained. In its first challenge of the post–cold war era, the United States established a coalition of thirty-eight countries to remove Iraqis from Kuwait after their August 1, 1990, invasion. Assembled as an overwhelming force, the U.S.-led coalition engaged in forty-three days of aerial operations, followed by a hundred hours of ground combat in Operation Desert Storm, totally routing Saddam Hussein's army and freeing Kuwait. Once again blacks played a role in every aspect of the liberation, with Gen. Colin Powell as the chairman of the Joint Chiefs of Staff and Lt. Gen. Cal Waller, a graduate of Prairie View A&M, serving as Desert Storm's second in command.

The national and world press focused on Powell as the senior U.S. military commander and the highest-ranking African American in the country's history. He disappointed no one. Appearing daily on global television and in newspapers, Powell displayed the intelligence, command presence, and overall leadership that soon had reporters referring to him as the "black Eisenhower" and speculating on his political future.

The son of Jamaican immigrants raised in the Bronx and a graduate of the City College of New York rather than West Point, Powell had served two combat tours in Vietnam and commanded at increas-

An aviation boatswain's mate lowers the flag on the USS *Lexington* on April 28, 1989, at Pensacola, Florida.
(U.S. NAVAL INSTITUTE)

ing higher levels as he advanced in rank. In 1987, Powell accepted President Ronald Reagan's appointment as national security adviser. Newspaper columnist Carl Rowan declared, "To understand the significance of Powell's elevation to this extremely difficult and demanding post, you must realize that only a generation ago it was an unwritten rule that in the foreign affairs field, blacks could serve only as ambassador to Liberia and minister to the Canary Islands."

After briefly returning to command all army troops in the continental United States, Powell again returned to Washington, this time as President George Bush's selection as chairman of the Joint Chiefs of Staff. Advances for African Americans had been slow for nearly two centuries, but in the period of one generation, an African American had risen on his own merit to command the world's strongest military force. Shortly after his selection, Powell remarked, "It was on this road to the future paved with the blood and sacrifice of black Americans that I became the first black chairman of the Joint Chiefs of Staff." He later added, "I wish that there were other activities in our society and in our nation that were as open as the military to upward mobility, to achievement, to allowing them in."

Unlike the operations in Grenada and Panama, the extended buildup period of Desert Storm provided more time for the American people and press to question the value and risks of such a commitment of U.S. manpower and resources. During the planning stages predictions about U.S. casualties ranged from a few thousand to thirty thousand, causing the military to order fifteen thousand body bags in anticipation of battle losses. Both black and white leaders recognized that while African Americans made up only 12 percent of the military-age people in the United States, 26 percent of troops in the Gulf were black. As a result, a *New York Times*/CBS poll of U.S. civilians revealed that 80 percent of whites favored action against Iraq, while only 50 percent of African Americans supported the impending war.

In a *USA Today* article on January 9, 1991, Mark Harrison, the national organizer of the African-American Network Against U.S. Intervention in the Gulf, questioned, "Why should we die for a country that has not given us equality?"

Most black leaders directed their protests against the government and the American society rather than the military. In a January 27, 1991, *Dallas Morning News* article, Lee Alcorn, president of a local NAACP chapter, wrote, "We feel it's unfair that the bulk of the fighting seems always to fall on minority groups, especially blacks and Hispanics. It's not necessarily an excuse to say it's an all-volunteer army. If we didn't have such a racist society, the army would not be an attraction to these people who can't otherwise realize the American dream."

Despite protests from groups outside the armed forces, blacks within the military stood united in support of their chain of command, remaining ready to follow lawful orders. In his autobiography General Powell recalled his response to questions about the number of African Americans in the war zone from black congressman Julian C. Dixon of California during the buildup in the Gulf. Powell wrote:

> I said I regretted that any American, black or white, might die in combat. But black fighting men and women, particularly in the all-volunteer force, would be offended to think that when duty called, they would be excluded on the basis of color. . . . The military had given African-Americans more equal opportunity than any other institution in American society, I pointed out. Naturally, they flocked to the armed forces. When we come before Congress saying we have to cut the forces, you complain we're reducing opportunities for blacks, I said. Now you're saying, yes, opportunities to get killed. But as soon as this crisis passes, you'll be back, worried about our cutting the force and closing off one of the best career fields for African-Americans. Do you want to have blacks in the military limited to the percent of blacks in the population, and throw the rest out? I don't think so. But you cannot have it both ways—favoring opportunity for blacks in the military in peacetime and exemption from risk for them in wartime. There was only one way to reduce the proportion of blacks in the military: let the rest of American society open its doors to African-Americans and give them the opportunity they now enjoyed in the armed forces.

The protests and controversy ended as quickly as the war itself. After less than five days of ground combat, an overwhelmed Iraq withdrew from Kuwait and agreed to a cease-fire. Coalition forces captured more than eighty-five thousand Iraqi soldiers in the brief conflict; an estimated eight thousand to fifteen thousand Iraqis died in combat.

American losses totaled 182 dead, 15 percent of whom were black. A Scripps-Howard News Service article, carried by newspapers around the country in early May, 1991, reported: "Another expert prediction fell victim to facts the other day: Expectations that black soldiers would do a heavy proportion of the dying in the Gulf War turned out to be wrong. . . . In any case, it's important to remember that the 182 American troops who died were black or white or Hispanic only secondarily. First and foremost, they were all Americans."

The United States emerged from the one-sided victory in the Gulf War as the preeminent world military power and as a political force that could unite the world against the aggression of a ruthless dictator. With the defeat of Saddam Hussein and no longer facing a global threat from the fragmented Soviet Union, the U.S. military resumed

First Marine Division members at the Kuwait International Airport on March 3, 1991, read about their success in liberating the country. (U.S. NAVAL INSTITUTE)

the reduction in forces that had begun before the invasion of Kuwait. Between 1989 and 1996 the army eliminated eight of its eighteen combat divisions, the air force cut the equivalent of twelve fighter wings, and the navy took ninety-four ships out of active service. Overall, the U.S. military reduced its manpower by 36 percent in seven years.

Despite these reductions, African Americans continue to be attracted to the military in numbers that exceed their proportion of the population. For the past fifteen years the percentages of African Americans in the armed forces has remained fairly consistent. At the beginning of 1996 blacks made up 21.9 percent of the total enlisted force—30.2 percent of the army, 16.8 percent of the air force, 18.5

percent of the navy, and 17.1 percent of the Marine Corps.

The number of black officers has substantially risen since 1981. In 1996, 7.6 percent of the total number of armed forces' officers were black—11.4 percent, army; 5.5 percent, air force; 5.4 percent, navy; and 5.6 percent, Marine Corps.

Black enrollment in the military academies increased, as did black graduates. The class of 1996 at West Point commissioned sixty-four black cadets, or 6.7 percent of the total. Enrollment in the academy's class of 2000 included eighty black cadets. Black midshipmen at the U.S. Naval Academy in 1996 made up 6.7 percent of the student body. The class of 1996 produced fifty-six black naval or marine officers. Seventy blacks reported in the 1,212-member Annapolis class of 2000. Air Force Academy statistics are similar, with blacks numbering 7.0 percent of the total enrollment, 70 of the 919 graduates of the class of 1996, and 79 of the 1,226 cadets reporting for the class of 2000.

Women, particularly blacks, have also thrived in the all-volunteer force. In 1996 women made up 12.5 percent of the total enlisted numbers and 13.0 percent of the officers. Black women constituted a large portion of those percentages. In the army 20.0 percent of the female officers and 47.9 percent of the enlisted women were black; in the air force 10.4 percent of the female officers and 24.7 percent of the enlisted women were black; in the navy 9.1 percent of the female officers and 29.1 percent of the enlisted women were black; and in the marines 7.6 percent of the officers and 25.3 percent of the enlisted force were black.

Many of the black female officers earned their commissions in the military academies. For the past five years the number of black women has averaged 15 to 20 percent of African-American graduates of the academies.

Recognition of the opportunities provided blacks by the military exists within and outside the armed forces. Edwin Dorn, of the Brookings Institute and later assistant secretary of defense for personnel and readiness, reported to the House Armed Services Committee on March 4, 1991:

> The military was the first major American institution to adopt and implement equal opportunity. Many of our civilian institutions have been slow to make such a commitment, and the results show in their

work forces, especially at the managerial and executive levels. The army had black officers commanding white soldiers in combat years before some of our universities entrusted black professors with the responsibility of teaching white undergraduates. It was not surprising that the Joint Chiefs of Staff had a black chairman before any Fortune 500 company had a black chief executive.

At a public hearing of the Defense Policy Panel on July 29, 1992, Les Aspin, chairman of the House Armed Services Committee, stated, "Today the military is one of, if not the, most thoroughly integrated institutions in American society."

Black civilian leaders also praised the military. In the March 1993 issue of the *Crisis*, James D. Williams, the public relations director of the NAACP, after a visit to U.S. bases in Germany, wrote: "Compared to other institutions, the military is miles ahead of them all in providing equal opportunity for African-Americans and other minorities. You did not need the statistics to show this. All it required was a look around on any of the installations to see the stripes on the sleeves and the bars, eagles, and stars on the shoulders, and to note who was giving the orders, to make this abundantly clear."

While the establishment of equality in the armed forces is a success story and the modern military far surpasses any other American institution in providing equal opportunity, racial problems nevertheless continue. The military still draws its members from a civilian society that does not offer the same equality, causing whites and blacks alike to arrive at induction centers with a lifetime of prejudices and opinions.

According to a poll of African Americans taken by *New Yorker* magazine that appeared in its April 29–May 6, 1996, issue, 58 percent of blacks say that their conditions in their civilian communities are getting worse. One-half of those polled stated that they believed that race relations would never be better than at the present time, and 59 percent agreed that the American dream has become impossible for most blacks to achieve.

A *Washington Post* survey published on July 15, 1996, revealed that 55 percent of the population in the South and 53 percent of the nation's population as a whole believe that racism is a "big problem" in American society today. Another 35 percent of both groupings agreed that racism is "somewhat of a problem." Only 3 percent of the two polling groups claimed that racism was "not a problem."

Difficulties between blacks and whites in the civilian communities still flare up into confrontations. With such conditions in civilian America, problems inevitably spill over into the armed forces. The occasional confrontation does occur, and latent racism and discrimination do continue.

A General Accounting Office (GAO) report released on November 20, 1995, revealed that African Americans were statistically less likely to gain promotions than their white counterparts in 80 of 116 measurements by service, rank, time in grade, and gender. According to the GAO, thirty-one of these deficiencies were statistically significant. The GAO, however, could come to no conclusion as to why the differences existed and admitted in its report that the disparities do not "necessarily mean they are the result of unwarranted or prohibited discrimination."

Blacks in the military still receive slightly more punishment than whites. According to an April 1995 GAO report, an April 1992 study by the Defense Equal Opportunity Management Institute concluded:

> There is a disparity in the judicial and nonjudicial punishment rates for black males. The punishment rates for both whites and blacks had been decreasing, but the black overrepresentation rate had increased slightly because the white punishment rate had been decreasing faster than the black punishment rate. Black males were overrepresented in the commission of violent and confrontational crimes, while whites committed the majority of crimes against property and military-specific offenses. Blacks were underrepresented in drug arrests and were overrepresented in courts-martial for crimes against persons.

None of the recent studies have investigated capital crimes, and the United States has not executed a military prisoner since 1961. However, it is noteworthy that in June 1996 six of the eight servicemen on death row at the Fort Leavenworth, Kansas, Disciplinary Barracks were black.

Despite its exemplary advancements in race relations, the military has not become complacent because of its successes. The Department of Defense still mans its equal-opportunity offices and continues race-relations training throughout the services. Each major change in the civilian and military leadership of the Department of Defense has produced an updated version of the Human Goals Charter, first published in 1969, and equal opportunity has remained an integral part of all manpower decisions. Between 1974 and 1994 the military conducted seventy-two studies of equal-opportunity issues.

One of the primary issues that continues to receive attention from African-American leaders and from the military itself involves black officers. In March 1994, Secretary of Defense William J. Perry directed a major study of the officer "pipeline" to increase the flow of minority officers into the services. Over the next year the services developed outreach, recruiting, and training procedures designed to increase the number of black officers in each of the services to 12 percent of the total by the year 2000.

The services have also been quick to react to actual and perceived increases in the number of hate groups and extremist organizations. After the bombing of the federal building in Oklahoma City, Secretary Perry and the chairman of the Joint Chiefs of Staff, Gen. John M. Shalikashvili issued a letter, on May 5, 1995, reaffirming the Department of Defense directive prohibiting military personnel from belonging to such organizations and encouraging commanders to employ all available assets to eliminate them from the service. According to the letter, "military personnel must reject participation in organizations that espouse supremacist causes; attempt to create illegal discrimination based on race, creed, color, sex, religion, or national origin; or advocate the use of force or violence, or otherwise engage in efforts to deprive individuals of their civil rights."

The military at all levels is well aware that its current status as the most integrated institution in American society did not come easily and that maintaining the position will require ongoing efforts. To ensure progress, the Department of Defense continuously issues reports, studies, directives, memorandums, and letters on race relations and monitors compliance with their instructions.

Secretary Perry best summarized these efforts and his demands for the services to maintain efforts to advance race relations in a March 3, 1994, letter: "Our nation's security and prosperity depend on our ability to develop and employ the talents of our diverse population. Equal opportunity is not just the right thing to do, it is also a military and economic necessity. . . . Therefore, I will not tolerate discrimination or harassment of or by any Department of Defense employee. The military services have led our nation in expanding opportunities for minority groups. . . . However, I believe we can and should do better on all fronts."

Perry's remarks tell the story of equal opportunity in the military today and the course it must follow in the future. Equality by birthright for African Americans has taken much too long, however, and the pride the military can take in its current race-relations posture is marred by centuries of racism and discrimination endured by African Americans in uniform. Fortunately for the military—and the defense structure of the country—there were African Americans willing to protest the status quo, break through racist barriers, and uphold the dignity of their race.

When Sgt. Maj. Edgar R. Huff, one of the first black marines, retired from the corps on September 28, 1972, after thirty years of service, he declared, "The Marine Corps has been good to me, and I feel I have been good to the Marine Corps."

More recently, army master sergeant Alex Pool wrote in a letter to the editor of *Army Times* that appeared in the July 15, 1991, edition:

> After 22 years, I am retiring. I am a Vietnam veteran and would not hesitate to go to war again if called upon.
>
> We do have quality blacks [in the army]. Just look at the pictures that came out of the Persian Gulf during the heat of battle. During the Vietnam War, how many blacks did you hear about going to Canada? We are a proud people and our color never runs.
>
> I did not enlist because I could not find a job. I enlisted because my brother enlisted before me. When he came home on leave with his starched khaki uniform, jump wings, and spit-shined jump boots, I knew I wanted to be a paratrooper.
>
> My mother had four sons; three served in the army. The brother I was so proud of was in Vietnam the same time I was. My mother did not ask why. She just prayed for our safe return.
>
> Sure, we still have bigotry in the armed forces. However, these same bigots came from the streets of America.
>
> Although I will miss the togetherness the Army instills in you to accomplish your mission, I have no regrets."

At a news conference on March 21, 1996, Togo West, the second African American to serve as secretary of the army said, "I want tomorrow the army we have today."

In his autobiography, Gen. Colin Powell concluded: "The Army was living the democratic idea ahead of the rest of America. Beginning in the fifties, less discrimination, a truer merit system, and leveler playing fields existed inside the gates of our military posts than in any Southern city hall or Northern corporation. The Army, therefore, made it easier for me to love my country, with all its flaws, and to serve her with all my heart."

For more than two hundred years African Americans have fought for their own personal freedom as well as that of their fellow Americans. Blacks contributed to the success of the revolution that gained the country, but not its slaves, their independence. Blacks played a significant role in preserving the Union in the Civil War and securing their own freedom. From the expanses of the American West to the heights of San Juan Hill, from the trenches of France to the heartland of Germany and Japan, from the icy mountain ridges of Korea to the thick jungles of Vietnam and the sands of the Persian Gulf, African Americans have performed loyally and bravely.

As the United States approaches the twenty-first century, African Americans serve in every specialty and in every rank in defense of their country. Their long struggle is not yet over, but African Americans have indeed proved that they, too, are Americans in every sense of the word.

BIBLIOGRAPHY

For the convenience of the reader, citations of periodical excerpts appear within the text of this book. Bibliographic materials herein listed for each chapter provide additional, more in-depth information on subjects of interest. Attributions to sources appear only in the first appropriate chapter even though the contents of the sources may extend into subsequent chapters.

Until recently, "mainstream" American historical studies ignored, omitted, or minimized the significant roles African Americans played in U.S. military history as well as other aspects of this country's development. Yet there exists a substantive body of work on specific, narrow subjects, such as the Buffalo Soldiers and the Tuskegee Airmen, much of it being written since the civil rights movements of the 1960s. The most detailed collection of black military references include the following:

Davis, Lenwood G., and George Hill. *Blacks in the American Armed Forces, 1776-1983: A Bibliography*. Westport, Conn.: Greenwood Press, 1985.

Slonaker, John. *The U.S. Army and the Negro: A Military History Research Collection Bibliography*. Carlisle Barracks, Pa.: U.S. Army Military History Institute, 1971.

Various general historical studies of black Americans in the United States proved useful in setting the scene and in providing the background against which military progress occurred. These include:

Ebony eds. *Ebony Pictorial History of Black America*, 3 vols. Chicago: Johnson Publishing Co., 1971.

Fishel, Leslie H. Jr., and Benjamin Quarles. *The Negro American: A Documentary History*. New York: William Morrow, 1967.

Foner, Philip S. *History of Black Americans: From Africa to the Emergence of the Cotton Kingdom*. Westport, Conn.: Greenwood Press, 1975.

Franklin, John Hope and Alfred A. Moss Jr. *From Slavery to Freedom: A History of Negro Americans*, 6th ed. New York: Knopf, 1988.

Hornsby, Alton Jr. *Chronology of African-American History: Significant Events and People from 1619 to the Present*. Detroit: Gale Research Inc., 1991.

Lerner, Gerda, ed. *Black Women in White America*. New York: Random House, 1972.

Comprehensive studies of African Americans in the U.S. armed forces are few, and they vary in merit. All, however, contribute information on black military history. Of particular value:

David, Jay, and Elaine Crane, eds. *The Black Soldier*. New York: William Morrow, 1971.

Farr, James Barker. *Black Odyssey: The Seafaring Traditions of Afro-Americans*. New York: Peter Lang, 1989.

Foner, Jack D. *Blacks and the Military in American History: A New Perspective*. New York: Praeger, 1974.

Greene, Robert E. *Black Defenders of America: 1775–1973*. Chicago: Johnson Publishing Co., 1974.

Lee, Irvin H. *Negro Medal of Honor Men*. New York: Dodd, Mead, 1967.

MacGregor, Morris J. *Integration of the Armed Forces: 1940–1965*. Washington, D.C.: Center of Military History, 1981.

MacGregor, Morris J., and Bernard C. Nalty. *Blacks in the United States Armed Forces: Basic Documents*. 13 vols. Wilmington, Del.: Scholarly Resources, 1977.

Moebs, Thomas Turxtun. *Black Soldiers—Black Sailors—Black Ink: Research Guide on African-Americans in U.S. Military History*. Chesapeake Bay, Md.: Moebs Publishing Co., 1994.

Mullen, Robert W. *Blacks in America's Wars*. New York: Pathfinder, 1973.

Nalty, Bernard C. *Strength for the Fight: A History of Black Americans in the Military*. New York: Free Press, 1986.

Office of Deputy Assistant Secretary of Defense for Equal Opportunity and Safety Policy. *Black Americans in Defense of Our Nation*. Washington, D.C.: U.S. Government Printing Office, 1985.

Shaw, Henry I., and Ralph W. Donnelly. *Blacks in the Marine Corps*. Washington, D.C.: U.S. Marine Corps, 1988.

Wilson, Joseph T. *The Black Phalanx: A History of the Negro Soldier of the United States in the Wars of 1775–1812, 1861–1865*. New York: Arno Press, 1977; originally published in 1890.

Chapter 1: Colonial Days and the Revolutionary War

Aptheker, Herbert. *The Negro in the American Revolution*. New York: International Publishers, 1940.

Moore, George H. *Historical Notes on the Employment of Negroes in the American Army of the Revolution*. New York: Charles T. Evans, 1862.

Nash, Gary B. *Red, White, and Black: The Peoples of Early America*. Englewood Cliffs, N.J.: Prentice-Hall, 1974.

Nell, William C. *The Colored Patriots of the American Revolution*. New York: Arno Press, 1968; originally published in 1855.

Quarles, Benjamin. *The Negro in the American Revolution*. Chapel Hill, N.C.: University of North Carolina Press, 1961.

Chapter 2: The War of 1812 and Beyond

Langley, Harold D. *Social Reform in the United States Navy: 1798–1862*. Urbana, Ill.: University of Illinois Press, 1967.

Laumer, Frank. *Dade's Last Command*. Gainsville, Fla.: University Press of Florida, 1995.

McConnell, Roland. *Negro Troops of Antebellum Louisiana: A History of the Free Men of Color*. Baton Rouge, La.: Louisiana State University Press, 1968.

Chapter 3: The American Civil War

Cornish, Dudley T. *The Sable Arm: Negro Troops in the Union Army*. New York: W. W. Norton, 1966.

Gallman, J. Matthew. *The North Fights the Civil War: The Home Front*. Chicago: Ivan R. Dee, 1994.

Glatthaar, Joseph T. *Forged in Battle: The Civil War Alliance of Black Soldiers and White Officers*. New York: Free Press, 1990.

Higginson, Thomas Wentworth. *Army Life in a Black Regiment*. Boston: Fields, Osgood, 1870.

Mays, Joe H. *Black Americans and Their Contributions Toward Union Victory in the American Civil War, 1861–1865*. Lanham, Md.: University Press of America, 1984.

McPherson, James M. *The Negro's Civil War: How American Negroes Felt and Acted During the War for the Union*. New York: Random House, 1965.

Quarles, Benjamin. *The Negro in the Civil War*. Boston: Little, Brown, 1969.

Uya, Okon E. *From Slavery to Public Service: Robert Smalls, 1839–1915*. New York: Oxford University Press, 1971.

Chapter 4: Reconstruction and the Indian Wars

Carrol, John M., ed. *The Black Military Experience in the American West*. New York: Liveright, 1973.

Fowler, Arlen L. *The Black Infantry in the West: 1869–1891*. Westport, Conn.: Greenwood Press, 1971.

Leckie, William H. *The Buffalo Soldiers: A Narrative of the Negro Cavalry in the West*. Norman, Okla.: University of Oklahoma Press, 1967.

Schubert, Frank N., ed. *On the Trail of the Buffalo Soldier: Biographies of African-Americans in the U.S. Army, 1866–1917*. Wilmington, Del.: Scholarly Resources, 1995.

Woodward, C. Vann. *The Strange Career of Jim Crow*. New York: Oxford University Press, 1966.

Chapter 5: The Spanish-American War: Cuba and the Philippines

Cashin, Herschel V., et al. *Under Fire With the Tenth U.S. Cavalry*. Salem, N.H.: Ayer, 1991; originally published in 1899.
Fletcher, Marvin. *The Black Soldier and Officer in the United States Army*. Columbia, Mo.: University of Missouri Press, 1974.
Gatewood, Willard B. Jr., ed. *Smoked Yankees and the Struggle for Empire: Letters From Negro Soldiers, 1898–1902*. Fayetteville, Ark.: University of Arkansas Press, 1987.
Lynk, Myles V. *The Black Troopers or the Daring Heroism of the Negro Soldier in the Spanish-American War*. New York: AMS, 1971.

Chapter 6: The Battles of Peace: Brownsville and the Great White Fleet

Harrod, Frederick S. *Manning the New Navy: The Development of a Modern Naval Enlisted Force*. Westport, Conn.: Greenwood Press, 1978.
Hart, Robert A. *The Great White Fleet: Its Voyage Around the World, 1907–1909*. Boston: Little, Brown, 1965.
Karsten, Peter. *The Naval Aristocracy: The Golden Age of Annapolis and the Emergence of Modern American Navalism*. New York: Free Press, 1972.
Lane, Ann J. *The Brownsville Affair: National Crisis and Black Reaction*. Port Washington, N.Y.: National University Publications, 1971.
Weaver, John D. *The Brownsville Raid*. New York: W. W. Norton, 1970.

Chapter 7: Prelude to World War: Mexico and Houston

Clendenen, Clarence C. *Blood on the Border: The United States Army and the Mexican Irregulars*. New York: Macmillian, 1969.
Haynes, Robert V. *A Night of Violence: The Houston Riot of 1917*. Baton Rouge, La.: Louisiana State University Press, 1976.
Whurfield, H. B. *10th Cavalry and Border Fights*. El Cajon, Calif.: Col. H. B. Whurfield, 1965.

Chapter 8: The Great War: World War I

Barbeau, Arthur E., and Florette Henri. *The Unknown Soldiers: Black American Troops in World War I*. Philadelphia: Temple University Press, 1974.
Bullard, Robert L. *Personalities and Reminiscences of the War*. Garden City, N.Y.: Doubleday, Page, 1925.

Carisella, P. J., and James Ryan. *The Black Swallow of Death: The Incredible Story of Eugene Jacques Bullard, the World's First Black Combat Aviator*. New York: Marlborough House, 1972.
Heywood, Chester D. *Negro Troops in the World War: The Story of the 371st Infantry*. Worcester, Mass.: Commonwealth Press, 1928.
Little, Arthur W. *From Harlem to the Rhine: The Story of New York's Colored Volunteers*. New York: Covici, 1936.
Scott, Emmett. *The American Negro in World War I*. Washington, D.C.: War Department, 1919.

Chapter 9: Between the Wars: The Struggle Continues

Dalfiume, Richard M. *Desegregation of the U.S. Armed Forces: Fighting on Two Fronts*. Columbia, Mo.: University of Missouri Press, 1969.
Patton, Gerald W. *War and Race: The Black Officer in the American Military, 1915–1941*. Westport, Conn.: Greenwood Press, 1981.

Chapter 10: World War II: Army

Arnold, Thomas St. John. *Buffalo Soldiers: The 92nd Infantry Division and Reinforcements in World War II, 1942–1945*. Manhattan, Kans.: Sunflower University Press, 1990.
Biggs, Bradley. *The Triple Nickels*. Hamden, Conn.: Archon Books, 1986.
Hargrove, Hondon B. *Buffalo Soldiers in Italy: Black Americans in World War II*. Jefferson, N.C.: McFarland, 1985.
Lee, Ulysses. *U.S. Army in World War II: Special Studies: The Employment of Negro Troops*. Washington, D.C.: U.S. Government Printing Office, 1966.
Luszki, Walter A. *A Rape of Justice: MacArthur and the New Guinea Hangings*. Lanham, Md.: Madison Books, 1991.
McGuire, Phillip. *He, Too, Spoke for Democracy: Judge Hastie, World War II, and the Black Soldier*. Westport, Conn.: Greenwood Press, 1988.
______. McGuire, Phillip. *Taps for a Jim Crow Army: Letters From Black Soldiers in World War II*. Lexington, Ky.: University Press of Kentucky, 1993.
Moore, Brenda L. *To Serve My Country, to Serve My Race: The Story of the Only African-American WACs Stationed Overseas During World War II*. New York: New York University Press, 1996.
Motley, Mary Penick. *The Invisible Soldier: The Experience of the Black Soldier, World War II*. Detroit: Wayne State University Press, 1987.

Chapter 11: World War II: Army Air Force

Davis, Benjamin O. Jr. *Benjamin O. Davis Jr., American: An Autobiography*. Washington, D.C.: Smithsonian Institution Press, 1991.
Francis, Charles E. *The Tuskegee Airmen: The Story of the Negro in the United States Air Force*. Boston: Bruce Humphries, 1955.

Osur, Alan M. *Blacks in the Army Air Forces During World War II: The Problem of Race Relations*. Washington, D.C.: U.S. Government Printing Office, 1977.

Sandler, Stanley. *Segregated Skies: All-Black Combat Squadrons of World War II*. Washington, D.C.: Smithsonian Institution Press, 1992.

Chapter 12: World War II: Navy and Marine Corps

Kelly, Mary Pat. *Proudly We Served: The Men of the USS Mason*. Annapolis, Md.: Naval Institute Press, 1995.

Nalty, Bernard C. *The Right to Fight: African-American Marines in World War II*. Washington, D.C.: Marine Corps Historical Center, 1995.

Purdon, Eric. *Black Company: The Story of Subchaser 1264*. Washington, D.C.: Henry B. Luce, 1972.

Stillwell, Paul, ed. *The Golden Thirteen: Recollections of the First Black Naval Officers*. Annapolis, Md.: Naval Institute Press, 1992.

Chapter 13: Executive Order 9981: Before and After

Gropman, Alan L. *The Air Force Integrates, 1945–1964*. Washington, D.C.: U.S. Government Printing Office, 1978.

MacGregor, Morris J. *Integration of the Armed Forces: 1940–1965*. Washington, D.C.: U.S. Government Printing Office, 1981.

Nelson, Dennis D. *The Integration of the Negro Into the United States Navy*. New York: Farrar, Straus and Young, 1951.

Stillman, Richard J. *Integration of the Negro in the United States Armed Forces*. New York: Praeger, 1968.

Chapter 14: The Korean War: Integration Under Fire

Milton, H. S., ed. *The Utilization of Negro Manpower in the Army*. Chevy Chase, Md.: Johns Hopkins University Press, 1955.

Rishell, Lyle. *With a Black Platoon in Combat: A Year in Korea*. College Station, Tex.: Texas A&M University Press, 1993.

Chapter 15: The Fifties: An Integrated Military in a Segregated America

Nichols, Lee. *Breakthrough on the Color Front*. New York: Random House, 1954.

Powell, Colin. *My American Journey*. New York: Random House, 1995.

Chapter 16: Vietnam: "The Only War We Had"

Goff, Stanley, and Robert Sanders. *Brothers: Black Soldiers in the Nam*. Novato, Calif.: Presidio Press, 1982.

Terry, Wallace. *Bloods: An Oral History of the Vietnam War by Black Veterans*. New York: Random House, 1984.
Westmoreland, William C. *A Soldier Reports*. New York: Doubleday, 1976.

Chapter 17: After Vietnam: Unrest and the Advance of Equality

Binkin, Martin, et al. *Blacks and the Military*. Washington, D.C.: Brookings Institution, 1982.
Lanning, Michael Lee. *The Battles of Peace*. New York: Ivy Books, 1992.
Moskos, Charles C. Jr., *The American Enlisted Man: The Rank and File in Today's Military*. New York: Russell Sage Foundation, 1970.

Chapter 18: United Defenders: Today and Tomorrow

Department of Defense. *Defense 95*. Washington, D.C.: U.S. Government Printing Office, 1995.
______. *Equal Opportunity Digest*. Washington, D.C.: U.S. Government Printing Office, 1996.
U.S. General Accounting Office. *Equal Opportunity: DoD Studies on Discrimination in the Military*. Washington, D.C.: U.S. General Accounting Office, 1995.
______. *Military Equal Opportunity: Certain Trends in Racial and Gender Data May Warrant Further Analysis*. Washington, D.C.: U.S. General Accounting Office, 1995.

INDEX

ABOUT THE AUTHOR

Michael Lee Lanning is a decorated twenty-year U.S. Army veteran who served as an infantry platoon leader and company commander in the Vietnam War. He is the author of ten books on military subjects, including *Vietnam at the Movies*, *Senseless Secrets: The Failures of U.S. Military Intelligence From George Washington to the Present*, and *The Military 100: A Ranking of the Most Influential Military Leaders of All Time*. Lanning resides in Phoenix, Arizona.